GDPR and Cyber Security for Business Information Systems

RIVER PUBLISHERS SERIES IN SECURITY AND DIGITAL FORENSICS

Series Editors:

WILLIAM J. BUCHANAN
Edinburgh Napier University, UK

ANAND R. PRASAD
NEC, Japan

Indexing: all books published in this series are submitted to the Web of Science Book Citation Index (BkCI), to CrossRef and to Google Scholar.

The "River Publishers Series in Security and Digital Forensics" is a series of comprehensive academic and professional books which focus on the theory and applications of Cyber Security, including Data Security, Mobile and Network Security, Cryptography and Digital Forensics. Topics in Prevention and Threat Management are also included in the scope of the book series, as are general business Standards in this domain.

Books published in the series include research monographs, edited volumes, handbooks and textbooks. The books provide professionals, researchers, educators, and advanced students in the field with an invaluable insight into the latest research and developments.

Topics covered in the series include, but are by no means restricted to the following:

- Cyber Security
- Digital Forensics
- Cryptography
- Blockchain
- IoT Security
- Network Security
- Mobile Security
- Data and App Security
- Threat Management
- Standardization
- Privacy
- Software Security
- Hardware Security

For a list of other books in this series, visit www.riverpublishers.com

GDPR and Cyber Security for Business Information Systems

Antoni Gobeo
Edinburgh Napier University, UK

Connor Fowler
Edinburgh Napier University, UK

William J. Buchanan
Edinburgh Napier University, UK

Published, sold and distributed by:
River Publishers
Alsbjergvej 10
9260 Gistrup
Denmark

River Publishers
Lange Geer 44
2611 PW Delft
The Netherlands

Tel.: +45369953197
www.riverpublishers.com

ISBN: 978-87-93609-13-6 (Hardback)
978-87-93609-12-9 (Ebook)

Contents

PART THREE: IMPLEMENTATION

Preface

Data misuse in the modern age has reached critical levels, without the public at large knowing the real extent the data they produce is being abused. The recent trends of data breaches from Equifax (2017) to companies such as Facebook and Google covertly selling personal data to analytics companies without the users consent, has prompted major shifts in the regulatory environment. The General Data Protection Regulation (GDPR) is the latest in data protection law to come out of the European Union. The regulation has created an increased awareness of the wild west nature of the data world, and how hidden and known players, malicious or otherwise, are attempting to control our worlds. This book was written, among other reasons, to allow the new generations of students to enter the business world with a greater sense of confidence in the face of ever evolving challenges in an increasingly connected world. The readers of this book should come away with a better understanding of the importance the GDPR has on shaping the digital and business world, and practical skills with which to apply that understanding.

The book is broken into three parts the introductory chapters (1–4), the preparation (5–7), and finally the implementation chapters (8–10). It is important that the reader reads part one first in order to gain a foundation of the basic subjects covered in the book (ideally cover to cover would be best). After having read part one there is no particular order in which the chapters of each part (2 & 3) should be read, as the principles covered in part one should give the reader enough knowledge to understand the topics covered in the respective chapters. It is not necessary to have read the GDPR from start to finish to understand the material covered in this book, the text was written in a way to simplify complex legalese into something that can be easily understood by students in their first year of University. The Articles most relevant to business information systems are quoted in the text so there is no need to have a copy of the regulation next to you as you read, but for the extra hard working, the full text of the regulation can be found online, for free, if you wish to read the Articles that are not mentioned in full detail. The book should also be able to be read in a way where one can easily pick it up to see, for example, how a business should conduct a DPIA? The answer to which can be learned if you read on.

For many people, using the internet is a common and daily occurrence, but they rarely see or understand the consequences of an untrustworthy internet, one that we have allowed to continue. It is our duty as authors to disseminate information of progressive moves, such as this regulation, towards building a society where the individual's digital and physical rights are held above the commercial interests of large corporations and governments. Together we hope to achieve that goal, thank you for reading.

Acknowledgements

It is a great gift to have the opportunity of working with such an inspiring educator as Bill Buchanan, without whom this book would not have been written; thank you Bill!

Thank you to Edinburgh Napier University, whose commitment to helping students achieve their highest potential has been on clear and constant display. A special thanks to the lecturers in the Business School, Faculty of Law, and to Mr Ken Dale-Risk; who patiently listened to many questions over the years.

There are key moments in life where someone else encourages us to believe in ourselves. My college lecturer, Ms Alison Bruce, urged me to have confidence in my abilities and helped me stay the course. Thank you, Alison!

To Anna and Dave: Thank you for your patient, stable, and wise support.

Most importantly; Tiberius and Freya, who endured many months of data protection "chats" and ready meals: you guys are Awesome! Thank you!

– Antoni

It has been a great experience diving into the world of data protection, but none of this could have been possible without the people in my life.

I would like to start by thanking the most important people in my life my love Antoni, who joined me on this journey, and the two brightest stars in my life Tibs and Freya.

A special thanks to Bill and Napier University who presented me with this great opportunity and John, the best cheerleader you could ask for, we should do dinner!

And finally, *per la mia nonna che ha cucinato sempre con amore, Dio ti benedica.*

– Connor

List of Figures

List of Abbreviations

AWS3	Amazon Web Simple Storage Service
BCM	Business Continuity Management
BIS	Business Information Systems
CEU	Council of the European Union
CIA	Confidentiality, Integrity, Availability
CIO	Chief Information Officer
CISO	Chief Information Security Officer
CJEU	Court of Justice of the European Union
CSIRT	Computer Security Incidence Response Team
CSO	Composite Signals Organisation
C-Suite	"Chief" managers in an organisation
DC	Data Controller
DCMS	Department for Digital, Culture, Media, and Sport
DP	Data Processor
DP	Data Protection
DPIA	Data Protection Impact Assessment
DPO	Data Protection Officer
DSAR	Data Subject Access Request
EC	European Commission
ECHR	European Court of Human Rights
ECJ	European Court of Justice
EDPB	European Data Protection Board
EDPS	European Data Protection Supervisor
ENISA	European Union Network and Information Security Agency
EP	European Parliament
FTC	Federal Trade Commission (USA)
GCHQ	Government Communications Headquarters
GDPR	General Data Protection Regulation
HMRC	Her Majesty's Revenue and Customs (UK)
IM	Information Management
InfoSec	Information Security
IOCCO	Interception of Communications Commissioner's Office

IoF	Institute of Fundraising
IoT	Internet of Things
IP address	internet protocol address
IPCO	Investigatory Powers Commissioner's Office
IPT	Investigatory Powers Tribunal
IRP	Incidence Response Plan
IS	Information Systems
ISACA	Information Systems Audit and Control Association
ISMS	Information Security Management Systems
ISO	International Standards Organisation
IT	Information Technology
JPCERT	Japan Computer Emergency Response Team Coordination Centre
MBR	Master Boot Record
MI5	Security Service (UK)
MI6	Secret Intelligence Service (UK)
MIS	Management Information Systems
NCA	National Crime Agency
NCSC	National Cyber Security Centre
NDA	Non-disclosure Agreement
NFP	Not-for-Profit Organisation
NGFW	Next-Generation Firewall
NGO	Non-governmental Organisation
NIS	Network and Information Security
NIST	National Institute of Standards and Technology (USA)
NSA	National Security Agency (USA)
OECD	Organisation for Economic Development
OODA Loop	Observe, Orient, Decide, Act Loop
OSC	Office of Surveillance Commissioners
PA	Public Authority
PaP	Policies and Procedures
PbD	Privacy by Design
PDCA	Plan, Do, Check, Act cycle
PII	Personally Identifiable Information
PLP	Principle of Least Privilege
QM	Quality Management
RM	Risk Management
SA	Supervisory Authority
SIA	Secret Intelligence Agencies
SIEM	Security Incidence Events Management
SIGINT	Signals Intelligence

1

Part One: Introduction

Chapters Covered

1) GDPR Fundamentals
2) Organisations, Institutions, and Supervisory Authorities
3) Business Information Systems
4) The GDPR and Cyber Security

Section Overview

This part of the book is intended to give the readers introductory knowledge of the subjects covered in this book. The topics covered will start with an analysis into the fundamental of the GDPR (the rights and principles), an overview of the supervisory authorities of the GDPR and the institutions and organisations affected by the legislation, what is business information systems and how they relate, and finally cyber security and how it can be utilised in the context of the GDPR. It will be helpful to think about this section through the lens of both and organisation and as an individual, allowing for a wholesome perspective on what the spirit of the legislation means. Below are some examples of things to consider.

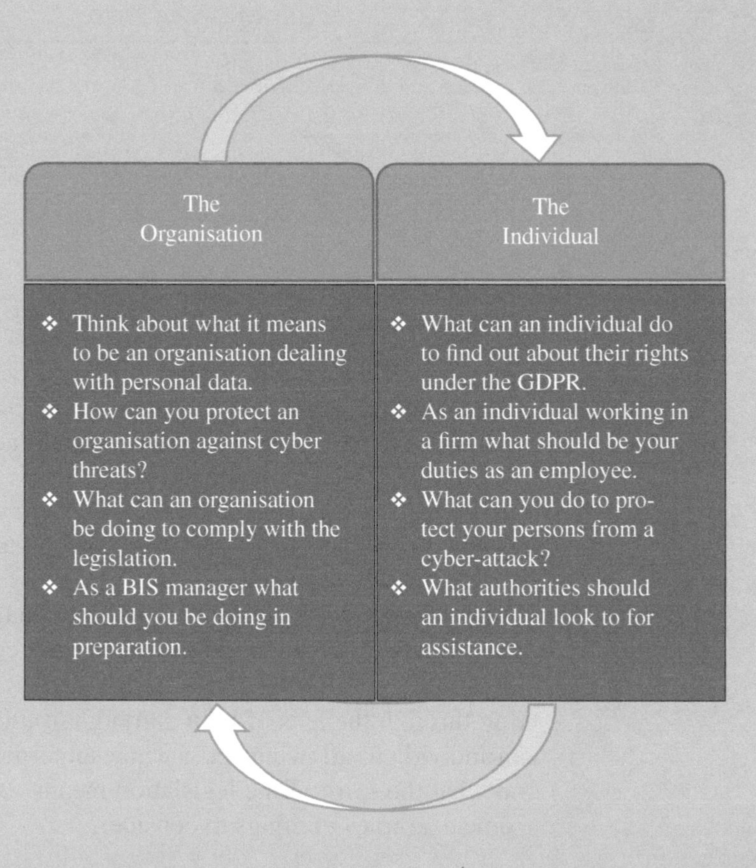
The Organisations and You
The Organisation
The Individual
❖ Think about what it means to be an organisation dealing with personal data.
❖ How can you protect an organisation against cyber threats?
❖ What can an organisation be doing to comply with the legislation.
❖ As a BIS manager what should you be doing in preparation.
❖ What can an individual do to find out about their rights under the GDPR.
❖ As an individual working in a firm what should be your duties as an employee.
❖ What can you do to protect your persons from a cyber-attack?
❖ What authorities should an individual look to for assistance.

Chapter 1: The GDPR Fundamentals

At a Glance:

- History of data protection and collection
- Personal data and its worth
- Rights of the natural persons under the GPDR
- Six Principles of the GDRP

Case Study: Cambridge Analytica

Learning Outcomes: *Students should be able to…*

- Understand the six principles underpinning the GDPR and their relevance in legislative compliance.
- Describe the rights of the data subjects and when and how they apply.
- Explain the value and uses of personal data and the potential consequences to the individual of its misuse.

Key Terms

1) Natural Persons
2) Personal Data
3) Data Subject
4) Data Controller
5) Data Processor
6) Rights and Principles
7) Data Minimisation
8) Lawful Basis

A Brief History of Data Collection and Data Protection

Governments and Institutions have been collecting personal data on their citizens since the beginning of recorded history. During the times of the Roman Empire taxation records were kept including the names, addresses, and incomes of Roman citizens. These records were consolidated and used for various purposes, depending on the desires of the emperor at the time. One thousand years later the Domesday Book of 1086 AD was instigated by William the Conqueror, in an attempt to clarify the rights to property and assets after the Norman conquest of England and Wales. It was the greatest survey of a nation's people and assets ever undertaken in Europe until that time, and the personal data collected was used for taxation purposes.

The interception of personal correspondence in "the national interest" has a long and royal pedigree. In 1516, during the reign of Henry VIII of England, the first "Master of the Posts" was appointed. This early form of postal service delivered royal mail; quite literally the mail for the king and his court. The convenience offered by this new service was offset by the knowledge that the letters would very likely be read by agents of the king. Later, in 1660 the General Post Office was established under the Reformation king Charles II, simultaneously establishing the "Secret Office" within the GPO, whose sole role was to spy on foreign correspondence entering and leaving England.

In modern times there has been an evolving recognition that individuals need to be protected from the misuse and abuse of their personal data, especially by governments and powerful organisations. In 1950, the European Convention of Human Rights enshrined the Right to Respect for Private and Family life (Article 8). This right includes the home and private correspondence as areas for protection within an individual's family and private life.

Building on this right the Council of Europe Convention 108 for the *Protection of Individuals with regard to Automatic Processing of Personal Data 1981* was introduced. This was the first internationally adopted law specific to data protection, with the clear aim of protecting the privacy of personal data. In 1995, the European Union Directive 95/46/EC on the protection of individuals regarding the processing of personal data and on the free movement of such data, was enacted. Known as the Data Protection Directive, it was the central piece of legislation on the protection of personal data in the EU and stipulated the requirement for explicit consent to the collection of personal data from the individual concerned.

The *Charter of Fundamental Rights of the European Union* included the protection of personal data as a fundamental right[1] and this is mirrored in the *Treaty on the Functioning of the European Union*[2]. Despite the clear intention to protect personal data, by 2012 it had become clear that the fragmentary nature of data protection legislation needed reform and the Commission of the EU published its suggestions. After three years of negotiations, the European Parliament and the Committee of the EU (made up of ambassadors from the 28-member states) finally agreed to the new regulation, and it became law in 2016.

The GDPR

The General Data Protection Regulation (GDPR) is the latest legislation on data protection arising from the European Union and came into effect in May 2018. It is designed to update the existing legislation to make it relevant to current technological trends and to bring conformity of compliance across the EU. There are three main objectives; to reinforce the protection of personal data for individuals, to assist the free flow of data within the single market (EU), and to reduce administrative burden. The situation within the EU Member states, prior to the introduction of the GDPR, was one of many fragmented and divergent sets of data protection laws relevant to each country. The harmonisation of data protection legislation across the EU allows for greater ease in the flow of data across the Union. It also allows for a specific authority to be nominated in each Member State; which creates a single point of reference for individuals and organisations. Member States may include specific additions to their local data protection laws, to provide further rights to their citizens.

Whilst similar to previous Data Protection legislation, the GDPR enhances the rights of Data Subjects and introduces enforceable new rights. Children are given a specific category of protection which acknowledges their vulnerability to the risks of sharing personal data; especially online. Children are also significantly less likely to know their rights regarding the processing of personal data.

In effect, the GDPR develops in further detail many of the already existing principles of data protection whilst encouraging a risk-based

[1] Article 8(1)
[2] Article 16

approach in assessing the level of protection required to meet the obligations for data controllers and processors to document their accountability and compliance.

This legislation is designed to deal with the ongoing challenges arising from the ever-changing online world and the massive increase in personal data volume and data flows globally. In response to increasing numbers of breaches of personal data, and to the lackadaisical attitude of many businesses, the new legislation has introduced significantly increased obligations and more onerous sanctions. The oft quoted ability to fine companies 4% of global turnover is only one example, and arguably not the most detrimental to the business.

To keep the regulation relevant in the face of unknown future developments, it is worded in technologically neutral terms. This means that the GDPR can be interpreted very broadly when decisions about what is within its scope, need to be made. Thus, it is extremely important to understand the objectives and principles that the GDPR is based on, as it will facilitate every step in the decision-making process.

To Whom Does It Apply?

The GDPR is primarily aimed at protecting data subjects. Data subjects are individuals who are identified or identifiable natural persons (living, breathing human beings) who are:

- Within the EU
- Whose behaviour takes place within the EU and is being monitored
- Natural persons whose data is being processed by an establishment within the EU.

The obligation to provide this protection also applies to organisations who are not based in the EU but who offer goods and services, with or without fee, to data subjects within the EU

Who Is Exempt?

There are two broad areas in which the GDPR does not apply; in the home, and for the purposes of law enforcement.

Household exemption

The processing of personal data by a natural person, for domestic or personal purposes, is not within the scope of the GDPR. The important aspects here are that the processing is by a natural person (an individual) and is in no way connected to professional or commercial activities. Therefore, doing your household budget or planning a family holiday is exempt from the requirements of the GDPR. However, work activities involving personal data that are carried out at home *are included* under the GDPR, for example an accountant working from home using client's personal data must demonstrate compliance to the regulation. This highlights the need to keep work and private/social activities separated; ideally on separate devices.

Law enforcement exemption

Under specific circumstances the competent public authorities are exempt from the requirements within the GDPR. This applies especially to those bodies engaged in the policing of, and sentencing for, criminal activities. Typically, this includes the security services, the police, and the judicial system. Other public authorities, such as the health services or taxation department for example, may be called upon to aid law enforcement. This exemption applies under the following specific circumstances:

- ***Criminal offences***: *In the prevention, detection, investigation, and prosecution of criminal offenses.*
- ***Criminal penalties***: *In the processing of personal data required for the application of criminal penalties.*
- ***Threats to public security***: *To safeguard the public against threats to security, and data processing for the prevention of such threats.*

Personal Data: Why it's Worth Protecting

Personal data under the GDPR is regarded as any personal information relating to a natural person; whether private, professional, or part of public life, it is considered as personal data. This data includes anything that could be used to personally identify an individual, either directly or indirectly, even where that data is considered to be generalised data. Some of the categories are: name, age, identification number, location, physical, mental, genetic, biometric, email, economic, social, historic, and more.

There is no comprehensive list of categories given within the GDPR; this is to allow any future developments in data gathering to be included within the scope of the regulation. This broad range of data sources becomes especially relevant where profiling of data is used. As you might imagine, personal data covering all these categories can provide comprehensive insight into a given individual. In some cases, this insight provides intimate knowledge about an individual; information that they, themselves, may not be aware of. This level of detailed information is commercially valuable, and this has given rise to a culture of data misuse and abuse within industries that commodify it.

The Privacy Argument

There is an overused refrain, which is often heard when the topic of data protection is broached: "I have nothing to hide, there's nothing interesting about me.", or conversely; "If you have nothing to hide, then you have nothing to fear." This implies that only "bad" people; criminals, terrorists, or other undesirables such as political dissidents, have something to fear from the analysis and the exposure of their private information. It also implies that the *only* people who can justify the desire for privacy are those same "bad" people. Another issue is that the notion of what constitutes "bad" behaviour is very narrowly defined by the public, but more broadly defined by governments and corporations. The public, in general, associates "bad" behaviour as that which damages an individual or society at large; behaviour that breaks fundamental laws, especially relating to the safety and security of the person and their property.

Institutions and corporations tend to see any behaviour that instigates unwanted change which affects them, as "bad". This is much broader in scope and may include; external criticism, competition, or challenges to their mode of operation for example, in 2018, the banning of cryptocurrency related discussion and advertising, on many social media sites such as YouTube and Facebook illustrate this, even though many governments around the world were researching the use or launch of a national cryptocurrency.

Similarly, this rhetoric gives rise to "self-surveillance". If I have nothing to hide, then my activities, especially online, must reflect my innocence, but do they? If I know that government agencies, social media sites, search engines, phone companies, and supermarkets etc. collect my data to use in ways I have little control over; I may change the way I live my life. When we know we are being watched, or think we are, it changes our behaviour. We become more compliant and conformist, rather than taking independent action based on our own agency. We act based on the perceived expectations of us, often based on societal norms and the desire to avoid shame and judgement.

After the revelations by Edward Snowden in 2013, which exposed the existence of the NSA/PRISM program and the extent of the mass surveillance of the public, researchers conducted a study into the rise of self-surveillance. They found that even in the absence of any government prosecution or punishment for individuals accessing certain "sensitive" information; in this case Wikipedia pages relating to terrorism, there was a steady and prolonged decline in the number of people searching for information on these subjects. The researchers concluded that the primary reason for the decline was the knowledge of the existence of mass surveillance by government agencies.

Ultimately, privacy is sacrificed by the individual in exchange for not being labelled as "bad". The dangers inherent in surveillance, through the collection and use of personal data, is mitigated only when the individual behaves in an unthreatening way. This is construed as complying with the status quo, or at the very least to keep dissent hidden and unspoken. This applies equally in politics as it does in commerce, and in the social sphere.

The Panopticon

The 18th century philosopher, Jeremy Bentham, developed the first modern system of mass surveillance and control; the Panopticon. A building designed to allow covert observation of all inmates held within it. The building was circular in design, with a central surveillance tower. The central tower allowed the guards to watch many inmates at once, but the most important aspect of the design was that the inmates could not see if, or when, they were being watched. Bentham realised that the outcome of this assumed surveillance was an increase in self-imposed conformist and compliant behaviour by the inmates. In 1843, he wrote that the panopticon was a new and unequalled method to gain control over another's mind. By the 20th century it was realised that this self-surveillance was the single greatest tool available for the mass control of human behaviour. In 1975, French philosopher Michel Foucault discussed self-surveillance as a a product of the "Disciplinary State", which strives to create a prison of the mind not of the body. He argued that self-surveillance is much more effective than punishment, enforced loyalty, or imprisonment, in the creation of a subjected citizenry.

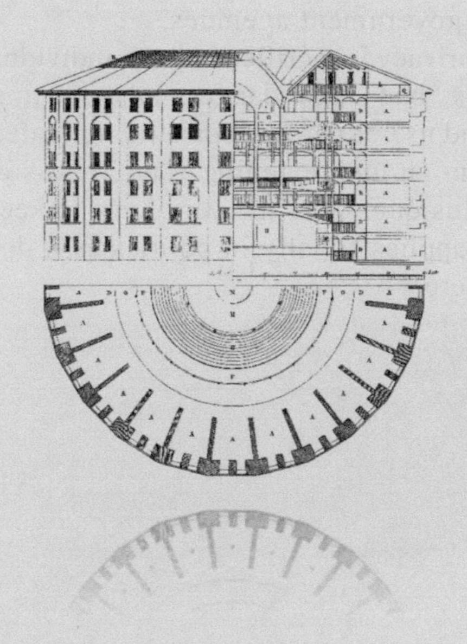

Image taken from: The Works of Jeremy Bentham Vol IV 172–3

The Economic Argument

You are worth it.

Every piece of personal data is worth something, somewhere, to someone, and as a result companies specialising in the "harvesting" of data to be processed for sale to third parties have flourished. The global sale of personal data to third parties was worth over £95 billion in 2012 and has been predicted to grow over 10% year on year. These "data brokers" use a wide array of techniques to collect and analyse personal data, most often without the knowledge or explicit consent of the individuals concerned.

Traditional marketing, known as offline marketing, using direct sales, print media, radio and television is distinguished by one key factor; the recipient is aware of its presence and has control over the level of engagement with the communication medium. Digital marketing using the internet, and especially social media sources, has begun to resemble covert surveillance. This is primarily due to the volume of data processed and the methods used in the collection and analysis of that data.

Julie Brill, former Commissioner of the US Federal Trade Commission (FTC), stated that these data harvesting companies maintain files identifiable to an individual. These files contain personal and specific information, including health data, family medical history, financial history and behaviour, sexual orientation, ethnicity and much more. They also have names or personal identifiers attached to

Cambridge Analytica: Case Study

"Today in the United States we have somewhere close to four or five thousand data points on every individual ... So, we model the personality of every adult across the United States, some 230 million people."

—Alexander Nix, chief executive of Cambridge Analytica, October 2016.

On March 19th, 2018, news broke about illegal data harvesting and manipulation by the data analytics company *Cambridge Analytica.* It emerged that the firm had been using data harvested from a third-party source, *Facebook.*

Facebook allowed the analytics company to process information on its users; data on their interests, dislikes, spending patterns, and other daily habits. The information was then used to sway undecided voters in various elections, such as the US elections of 2016.

The consequences of the data mismanagement cost *Facebook* an estimated £35 billion of market value and led to a US government inquiry.

The exposure of *Cambridge Analytica's* activities led to increased scrutiny of the company's business model, revealing it had consistently broken laws under the EU Data Protection Directive.

The Information Commissioner's Office, took legal action against the firm, obtaining a warrant to search the headquarters on March 23rd, 2018.

Shortly after, *Cambridge Analytica* ceased operating.

them and as such are used as a digital copy of an individual, which is sold on by the brokers as a product.

Consequences to Individuals of Data Misuse

Financial damage

Misuse of personal information has consequences to the individual concerned. The most obvious being identity theft and fraud leading to financial loss. In 2017 it was found that individuals are 35 times more likely to be victims of fraud and computer misuse, than of a robbery. In the same year, it was estimated that hackers stole approximately £130 billion from consumers worldwide. The victims of these types of crimes are often held financially accountable for the subsequent damage and loss. The local laws, both national and regional, determine the burden of proof required, and the level of protection offered in cases of financial crime. Misuse of data by third parties leaves individuals, whom have acted in good faith, vulnerable to attacks on their personal data.

In February 2018, the Amazon Simple Storage Service (AWS3) was found to contain the unsecured personal data of over 119,000 Fedex customers. Fedex is a global parcel delivery company. This data was open to anyone, and contained the scanned images of passports, driver's licenses, and security documents of customers worldwide. Also unsecured were the; names, home addresses, phone numbers and other personal information of customers.

How many records were copied, for what purpose, and by whom, was unknown; however, the consequences of identity theft often arise only when the victim applies for a new loan or receives the results of a credit check or similar. This means that the impact may be felt by the victims for many years. Considering that identity fraud for financial gain carries a risk to the victim of unclearable debt, damaged credit rating, unpaid bills etc. the clear disregard shown by FedEx, for the consequences of data misuse on the lives of their customers, is a compelling reason for increased accountability.

Reputational and psychological damage

Data misuse can harm individuals in non-physical ways. Private information which is exposed to third parties, either specifically or *en-masse*, can have devastating consequences to the individual's reputation and psychological well-being. This type of damages has been recognized by the courts as a civil wrong (delict/tort) which may be actioned in a claim for damages or compensation.

The case of *Vidal-Hall & Othrs v Google*, established that distress suffered because of the misuse of personal information was valid grounds in a claim for damages. The case was based on the collection, by Google, of the claimants' Browser-Generated Information (BGI) without their knowledge or consent. This was achieved using "cookies" and was processed and sold to third parties; advertisers, who used the information to display advertising targeted to the specific individual. The case was brought claiming the advertising displays could have been, and were, seen by third parties and that this revealed private information about the individuals concerned. It was argued that the open display of these advertisements damaged their personal dignity, integrity, and autonomy, and the Court agreed. Prior to this case, individuals were required to show that they had suffered an actual loss; that is a quantifiable financial or material loss.

Processing, of data, especially for marketing purposes, may expose personal information about an individual as a by-product of the original communication, and as such would constitute a breach of their privacy.

Life and livelihood

Personal data acts as a reflection of an individual's life and livelihood. The potential of this information, to be misused and affect their outer world, can have grave consequences.

Data misuse may have a direct and negative effect on the individual concerned, one that has physical or material consequences. Misuse of health data may lead to incorrect treatment, or diagnosis. In some countries, employers may use personal health data, including mental health data, to inform their decisions on new, or continued employment opportunities offered. Insurance companies rely on accurate personal data to assess the level of insurance cover they are willing to offer an individual. Mortgages are offered based on data which references the above-mentioned areas of health, employment, and insurance.

As an example, the personal data available on social media sites, such as LinkedIn or Facebook, may be misused by potential employers when assessing candidates for work.

The Heart of the GDPR; The Six Principles

The six principles of the GDPR are the guiding ethical intentions underpinning the legislation. The principles set forth the vision for data protection going forward. It outlines the ethos of the legislation and the people who are accountable in upholding its values and obligations. Aligning oneself to the six principles of the GDPR and imbedding them throughout the organisation, will ensure a sound foundation in meeting its requirements. Using them as the basis for all decision making relevant to the processing of personal data within organisation creates a deeper level of understanding of the spirit of the law.

Lawfulness, Fairness, Transparency

Principle #1 (Art 6, 5(2), 24, Rec 171, 40)

The GDPR requires that all personal data processing must be done in a lawful, fair, and transparent manner. This means that the legality of the basis, that is the reason for the purpose of the processing, must be stated and documented in a clear and transparent way, and to which the data controller will be held accountable. This tripartite principle is extremely important as it underpins all the other principles and rights within the GDPR.

Lawfulness here means that the processing is being done under one of the six specific purposes named within the legislation. Any reason outside the six specified is considered to be unlawful.

Fairness gives individuals control and choice over what data is collected, how, if, or when it is processed, and why it should be collected and processed at all.

Transparency is achieved by keeping the individual informed of the collection, purpose, and processing of their personal data. The more complex the data, and the relative scale of the medium from which the data is drawn, the greater the need for clear and direct communication, which must be done at the point of collection and before any processing begins.

All communication must comply with the Right to be Informed, which provides for communications to be in clear, concise, transparent and easily intelligible language. Plain and clear language is especially relevant where communications are directed at children. There are six bases for lawful processing, at least one of which must be chosen before processing begins.

The Six Lawful Bases

Consent

Consent is the explicit and clear agreement, given by the individual, to process their personal data for a specific purpose. Consent must also be informed and unambiguous and the controller must be able to show that all these conditions for consent have been met. Thus, the use of pre-ticked boxes, inactivity, or silence do not constitute consent from the individual. The consent must be freely given and must not be a condition for the performance of a contract if the processing is not required to fulfil that same contract. Where the request for consent forms part of a written declaration, which covers multiple areas or subjects, then those parts of the declaration which infringe the GDPR will be considered as non-binding.

The consent applies only to the specific purpose stated, and where multiple purposes for processing exist, separate and specific consent must be given. Where sensitive data is to be processed, the specific consent for the processing of the sensitive personal data must be obtained from the individual. In the case of children, the holder of parental responsibility must give, or authorise the specific consent for the child. The GDPR regards children as being those under 16 years old, and in some cases 13 years old may be the age of consent. Under no circumstances can a child under 13 years give their consent without parental approval.

Contract

A contract is a legally binding agreement between two or more parties. The performance of a contract is the fulfilment of its bilateral obligations and duties. Where the data subject has requested that specific steps be taken before entering into a contract, then the processing of their data maybe legitimised for that specific purpose.

If an individual wanted to compare the cost between one energy supplier and another, he may request that is energy usage data be analysed before agreeing to a new contract. Were the processing is necessary for the performance of a contract between the data subject and the data controller; for example when a bank provides a mortgage, then this is a lawful purpose in contract.

Legal obligation

Within the GDPR, a legal obligation specifically refers to the laws of the European Union and the laws of its Member States. This purpose applies when the processing is necessary for compliance to these laws, as in the case of anti-fraud legislation where controllers may be required to pass certain categories of data to relevant government authorities.

Vital interests

Primarily referring to the preservation and protection of an individual's life, the vital interests purpose will generally apply in a medical context, although not exclusively. The processing must be an important aspect in the provision of the protection to their life. Emergency departments in hospitals are an example of a place where this purpose would be exercised.

Public task

Public tasks are undertaken by Public Authorities and this purpose, as a ground for data processing, may only be used by public authorities. What constitutes a public authority is discussed in chapter two. A public task is one which has a clear basis in law and it is required for the performance of a task in the public interest or for official functions. This purpose covers only those activities specific to the legally defined tasks of the public authority; those activities which allow it to fulfil its functions in the public interest. It does not include activities for marketing or research purposes, or any other purpose, out with the legally defined one for that specific public authority. All publicly owned organisations, government institutions and bodies, including military and health, must process data which meets the criteria under this purpose exclusively.

Legitimate interests

After consent, the legitimate interest purpose is the most broadly applicable to all organisations which are not considered to be public authorities. Despite its broad application, controllers must ascertain whether they do, in fact, have a legitimate purpose in processing personal data. It is not sufficient simply to establish that there is a business or commercial case for the processing; it must be shown that the processing is necessary and not simply convenient.

Where this is the case, the legitimate interest of the controller or of a third party will be lawful unless there is a good reason to protect the individual's fundamental rights and freedoms which overrides those legitimate interests. This is especially relevant where the data subject is a child. The legitimate interest purpose must be ascertained after an assessment that includes consideration of whether the individual might reasonably expect their data to be processed in this way. The context under which the data was collected, and the time at which it was collected should be factors in the assessment of what might be determined to be a reasonable expectation.

Data processing for direct marketing purposes is considered a legitimate interest. This cannot apply if you are a public authority processing data to perform your official tasks. Public authorities have specific legislation which defines their purpose in the processing of personal data.

Where no lawful basis can be attributed to the processing of personal data, any processing will be considered unlawful and the individual has the right to request the erasure of that data. The lawful basis for processing may also affect which other rights are available to the individual concerned, since not all the rights are absolute. It is important to analyse the reasons for the processing before deciding on the lawful basis as it is not designed to be easily changed. Rather, the principle of Lawfulness is designed to encourage Controllers and Processors to assess the legitimate reasons for data processing, and to minimise the occurrence of data processing outside the original transaction. The importance of a detailed and clear Privacy Notice stating the intended purpose or purposes of the processing and the legal basis for that processing will be discussed in detail in chapter 6.

Necessary processing of data, whilst not necessarily essential, must be achieved in a proportionate and targeted manner. It must be a necessary part of fulfilling the purpose stated, rather than simply the preferred method chosen to fulfil that purpose.

Purpose limitation

Principle #2

The GDPR clarifies the circumstances under which personal data may be processed by highlighting areas which have hitherto often been ignored. The principle of purpose limitation creates a stringent requirement limiting the way in which data may be processed once it has been collected. Personal data must only be processed for the original purposes for which it was collected. It must be collected for a specific and explicit purpose and that purpose must have a lawful basis which forms the only reason upon which that data may be processed.

Any change of purpose requires an assessment of the compatibility of the new reason for processing the data relevant to the original purpose. The new assessment must consider the fairness of the new purpose relevant to the individual. This is ascertained by considering the possible effect on the privacy of the individual and whether this new purpose would be a reasonable expectation, by the individual, for the use of their data.

Change of original purpose for processing

Where the original purpose for processing has changed, any continued processing of that data must be done under a new purpose. However, it is not just a matter of choosing another purpose from the list, as any change in purpose must be justified through an assessment. Where the original purpose had its lawful basis in consent, then the new purpose must have a different basis from that for which consent was originally given.

Consent must always be informed and specific to the process in question, therefore any changes to the purpose behind the processing of personal data must receive fresh consent. Where the new purpose of processing is compatible with the original purpose (other than consent), and this can be demonstrated, then continued processing may be possible.

To assess whether the new purpose is compatible with the original purpose any link between the original purpose and the new purpose should be established. The context under which data was collected, taking account of the relationship with the data subject and the reasonable expectation that they may have regarding its use, is fundamental in assessing compatibility of purpose. where special category data, or criminal offense data, is concerned the potential consequences for the data subjects must be explored and safeguards applied.

Data minimisation

Principle #3

Following on from the Purpose limitation principle, the personal data collected should be only that data which is adequate, relevant and limited to what is necessary for the specific purpose stated. Minimisation means collecting only that data which can and will be used for the purpose specified. Many businesses collect far more than is required in the belief that it may become useful one day. The GDPR requires controllers to demonstrate the need for the collection and processing of personal data, and where that need cannot be demonstrated they may be considered as non-compliant.

Accuracy

Principle #4

Personal data must be accurate and updated where necessary. Any inaccuracies, relevant to the purpose for the processing, must be rectified or erased as soon as possible; that is without delay. Accuracy of data relates to raw data collected from or about the individual and the method of analysis applied to that data. The need for accurate data is obvious, but it has greater implications in the processing of big data, especially when used for the profiling of individuals. Since profiling entails the creation of data, by inference or derivation, any inaccurate results may negatively impact the individual where it is used as a predictive tool. This is especially acute where it relates to health, finance, or behaviour, and as such may be inherently unfair.

Storage limitation

Principle #5

Data which allows the individual to be identified must be kept for no longer than is absolutely needed for the purpose for which it was collected. The only circumstances in which this data may be archived or stored is when it is used in the public interest, historical or scientific research, or certain statistical purposes. In all circumstances where the data is retained, the rights and freedoms of the individual must be safeguarded with the appropriate technical and organisational measures.

Integrity and Confidentiality

Principle #6

This principle seeks to ensure the security of personal data by protecting it against unauthorised, unlawful, or accidental processing, loss, destruction, or damage. An adjunct to this is the potential for unauthorised disclosure of personal data. Appropriate technical and organisational measures are required to provide this security and to demonstrate compliance. This is a key area in that it presents internal and external risk factors which must be assessed and mitigated using the measures appropriate to the level of risk.

At a minimum it is required under Art 32(1) Security of Processing) that:

a) The pseudonymisation and encryption of personal data
b) The ability to ensure the ongoing confidentiality, integrity, availability and resilience of processing systems and services;
c) The ability to restore the availability and access to personal data in a timely manner in the event of a physical of technical incident
d) A process for regularly testing, assessing and evaluating the effective ness of technical and organizational measures for ensuring the security of the processing

Article 32(1)

It is important to note that these requirements are to be applied as appropriate, and that the blanket encryption of personal data is not, in itself, sufficient. Following an approved Code of Conduct or certification is an additional means to show compliance. The relevant supervisory authority in each EU Member State is the primary source for information

Social Credit System: today, China; tomorrow the World?

Imagine a day where you're refused access to public transport because of your credit score. Big data and online activity are being used in China as measurements in a system for the; control, reward, and punishment of citizens. Access to schools, transport, and jobs are some of the rights controlled by the Social Credit System, this scheme extends to businesses and governmental bodies. By 2020, everyone in China will be ranked, and given a score, in at least one of the following four areas:

- Societal Integrity: for individuals.
- Commercial Integrity: for businesses operating within China.
- Judicial Credibility: for judges.
- Honesty in Government Affairs: for civil servants.

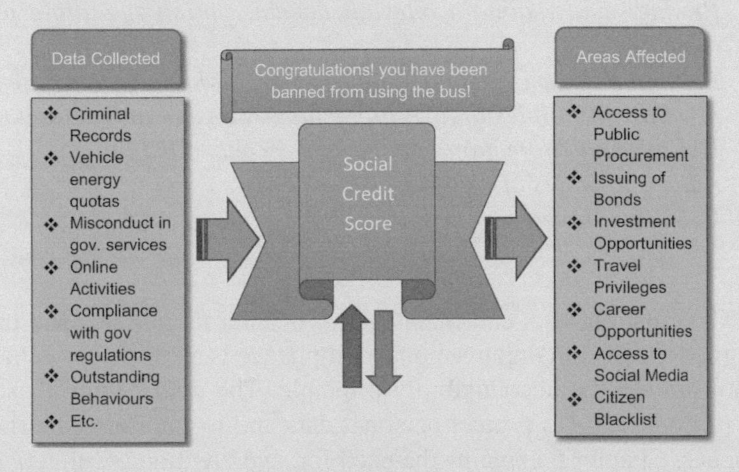

According to the Chinese government system has one primary goal, to engender trust, through the "raising of awareness of individual and group integrity, and the level of credibility within society". This is achieved through the awarding of a "reputation" score based on the analysis of big data. Private companies, such as Alibaba and Tencent, were involved in the initial stages of developing the Social Credit score, based on financial credit scoring models. Reputation systems, such as this one, already exist outside China: think of Amazon.com's rating system, or Ebay. The Chinese government uses data from governmental and commercial sources to evaluate and score individuals and others.

on conduct and certification, as each state has its own laws and regulations which apply.

Accountability and Governance

Although not considered one of the 6 key principles of the GDPR, the principle of Accountability is explicit within the regulation and requires compliance with all the other principles to be clearly demonstrated. The responsibility to demonstrate compliance is highlighted and it is expected that measures which are both comprehensive and proportionate will be adopted by the data Controller. There are specific measures which must be applied, these are:

- *Demonstrate appropriate technical and organisational measures.*
- *Produce and maintain relevant documentation regarding processing activities.*
- *Appoint a Data Protection Officer (DPO) where appropriate.*
- *Meet the principles of data protection by design and default.*
- *Use of Data Protection Impact Assessments (DPIA).*
- *Adhere to codes of conduct.*
- *Use of approved certification schemes.*

The creation, use, and maintenance of internal policies and procedures designed to assist in organisational compliance is a key tool in achieving and demonstrating the accountability principle. The goal behind these policies and procedures is to protect personal data and to minimise the risk of data breaches. Chapter 6 explains the need for, and creation of, effective policies and procedures in detail.

The Rights of Natural Persons in the GDPR

The GDPR describes the rights afforded to the individual in the new era of data. There are eight fundamental rights granted to natural persons under the GDPR. This section intends to conceptualize and develop a deeper understanding of the eight rights and their related articles and recitals.

The Right to be Informed

(Article 12, 13, 14)

The data subject must be informed of the collection and use of their data. This is part of the requirement for transparency within the GDPR, and especially relates to "privacy information". Privacy information encompasses the purpose behind the collection of the data, how long it will be held for, and any third parties with whom that data will be shared. This can be reduced to three questions: Why? How long? With whom?

This information must be given at the time the data is collected and must be communicated in a clear, and easily understood, simple and plain language. Of course, the language must be one that is intelligible to the data subject, that is their native language in the first instance. The information must also be easily accessible to the data subject. If the data is obtained through third parties, the same requirements apply, however there is scope for the data subject to be informed no later than one month from the time of collection of the data.

There is no need to notify the individual if they already have the information, or if doing so would involve disproportionate effort. Individuals must be notified if collected data is to be processed for a new use, for which they have not previously been notified. This must be done before any processing begins.

The Right of Access

(Article 15)

The Right of access allows individuals to receive confirmation regarding whether, or not, personal data about them is being processed. Where processing of their data is taking place, the controller must provide a copy of the data being processed without charge. Individuals have the right to access their personal data and any supplementary information relevant to that data. This allows them to be aware of the data collected and to verify the lawfulness of the collection of that information. This right follows directly from the right to be informed, in that its objective is to allow the individual to know that data is being collected on them, the purpose of the processing, and to check if the correct legal procedures have been followed.

The Right to Rectification

(Article 16, 12, 19)

Where information is inaccurate or incomplete, individuals have the right to request its rectification. The rectification must be done within one month of the request being made, or in cases of very large and complex requests the time period may be extended by another two months. The individual must be informed of the reasons for the extension and they must also be informed of their right to complain to the relevant supervisory body, and their right for judicial remedy.

If the inaccurate or incomplete data has been shared with third parties, these third parties must be individually contacted and informed of the rectification. There is an exception in cases where this would prove impossible or involve disproportionate effort. Individuals may request the names of these third parties, and where this is the case the names of the third parties must be shared with the individual.

The Right to Erasure

(Article 17, 19)

The objective of this right, sometimes erroneously referred to as the right to be forgotten, is to allow the individual to have their personal data removed or deleted where there is no ongoing and compelling reason for its continued processing or storage by the collector. Where the data has been made public it is required that reasonable steps be taken to inform other controllers of the request for erasure. This includes the deletion of links to, and copies or replications of, that data where it is technically and economically feasible. The right does not apply in cases where it is in the public interest by an official body, or where the Right to Freedom of expression and information is being exercised. The Right to Erasure is actionable in specific circumstances:

- *Where the personal data is no longer necessary in relation to the purpose for which it was originally collected/processed.*
- *When the individual withdraws consent.*
- *When the individual objects to the processing and there is no overriding legitimate interest for continuing the processing.*

- *The personal data was unlawfully processed (otherwise in breach of the GDPR).*
- *The personal data must be erased to comply with a legal obligation.*
- *The personal data is processed in relation to the offer of information society services to a child.*

The Right to Restrict Processing

(Article 18, 19)

Data subjects have the right to supress or block the processing of their personal data. This applies where the accuracy of the relevant data is contested and requires verification. Whilst the processing of the data is being contested, only the continued storing of the data, and no other processing, is permitted. If the data was collected in the public interest, or for legitimate purposes, the processing of that data must be restricted until the legitimate grounds for the continued processing of the data are established. There may be circumstances where the processing of the data itself is unlawful, however the individual data subject requests restriction rather than erasure. The data subject also has the right to request restriction where the data is no longer required by the collector/processor but the individual wishes to establish a legal claim based upon it.

Individuals must be informed if or when a restriction on the processing of their personal data has been lifted, and if they so request the names of any third parties with access to that data. Similarly, any third parties in receipt of, or with access to, the data must be informed of the restriction, unless this proves impossible or involves disproportionate effort.

The Right to Data Portability

(Article 20, 12)

The Right to data portability is designed to strengthen the individual's control over their personal data, and to encourage competition between data controllers. This right gives individuals control over where and to whom the access to their personal data is granted. It allows individuals to copy, move, or transfer their data across different service platforms and it provides for the safe, secure, and smooth provision of portability for the individual to obtain and reuse their data as best fits their requirements. The right applies specifically to the automated processing of data provided by the individual to a controller, based on consent or contract. The individual must specifically request for the data to be transmitted to a third party, and this must be executed where feasible. Where the data relates to more than one data subject then the impact of the portability of that data on the rights of those individuals must also be assessed.

A response to a request for data portability must be given without undue delay, within one month of receipt. If it is decided that no action will be taken, the individual must be notified as to the reason and they must be informed of their right to complain to the supervisory authority, and to take judicial remedy. The information must be provided without cost to the individual, and should be provided in a structured, and commonly used, machine readable format which enables other processors to extract relevant data.

The Right to Object

(Article 21, 12)

Where the processing of personal data is being carried out based on the "legitimate interest" purpose, or by an official authority, the data subject is entitled to object on the grounds of their particular situation. The right to object relates to the processing of data based on legitimate interests, or tasks performed in the public interest by official authorities, for direct marketing purposes, or for scientific and historical research. In all cases, including any automated profiling, the individual must be explicitly informed of this right at the first point of communication. All processing of the relevant data must cease unless it can be clearly and definitively shown that overriding and legitimate grounds for processing exist.

The processing of the data for legal claims is explicitly cited as one such ground. Similarly, processing for scientific or historical research may be objected to on the grounds of the individual's particular situation only. Where that research is done in the public interest, there is no requirement to comply. Objections relating to direct marketing must be actioned immediately, as there are no grounds for refusal or exception. All online variants of these categories must offer a means to object at the point of first communication.

Rights in Relation to Automated Decision Making and Profiling

(Article 22, 4(4), 9, 12, 13, 14, 15, 21, 35(1), 35(3))

Any form of automated processing of personal data consisting of the use of personal data to evaluate certain personal aspects relating to a natural person, in particular to analyse or predict aspects concerning that natural person's performance at work, economic situation, health, personal preferences, interests, reliability, behavior, location or movements.

Article 4(4)

This right is considered especially important within the GDPR, as can be seen by the number of articles relevant to it. It concerns decision making related to an individual, which is conducted solely through automation, without human input into the decision-making process, and includes profiling.

The legislators consider this type of data processing as high risk when it carries the potential to adversely affect an individual's legal rights or an area of similar significance such as finance or health. There are also further restrictions in the processing of children's data, and other special categories of data. Provision is made for exceptions, on three grounds, and where these apply additional safeguards must be implemented to further protect the relevant data.

The data subject shall have the right not to be subject to a decision based solely on automated processing, including profiling, which produces legal effects concerning him or her or similarly significantly affects him or her.

Article 22(1)

Individuals must be given specific information about the processing; such as the logic process used in the decision making. An explanation of the impact on the individual from the potential consequences arising from the process must also be provided. Individuals must be provided with all the following:

- be able to obtain human intervention in the decision-making process,
- be allowed an opportunity to express their opinion/point of view,
- receive an explanation of the decision and,
- have a means to challenge that decision if they wish to.

There is an obligation on the Controller to show that errors, bias, and discrimination are actively prevented since individuals also have the right to challenge the decision and request that it be reviewed. It is therefore important that the data on which the decision is made is processed using the most appropriate statistical and mathematical models. Concurrently, the use of best practise measures, both organisational and technological, must be demonstrated to ensure:

- minimisation,
- rapid identification and,
- correction of errors.

The personal data collected must also be held securely, in such a way as to reflect the level of risk posed to the individual's rights and freedoms. However, it must also be secured without giving rise to discriminatory effects for the individual.

The Three Exceptions

Article 22 allows this type of fully automated decision making to take place under three limited circumstances:

1. *If the process is necessary for the performance of a contract or entering into a contract between an organisation and an individual. This restriction may NOT be lifted in the processing of Special Category personal data.*
2. *Where the processing is authorised by European Union law, or the law of a Member State, and these laws apply to the data controller.*
3. *The processing is carried out with the individual's explicit consent.*

Where the processing does come under the remit of Article 22, then individuals must be informed of the processing, and easily accessible methods of requesting human intervention, or the ability to challenge a decision, must be made available to them. Regular system checks to monitor performance must also be demonstrated to show that they are working as intended. Special categories of data may be processed by automated means, however there must be explicit consent from the individual concerned or a *substantial* public interest must be demonstrated.

The planned processing of high risk data always requires a Data Protection Impact Assessment (DPIA) to be conducted. This document allows risks to be identified, assessed, and mitigation strategies to be outlined. It is an important step in showing compliance and is discussed in detail in chapter 7. Automated decision making, and profiling, may be carried on in circumstances where the data does not carry legal or similarly significant consequences for the individual; however, the lawful basis for such processing must be documented. Individuals have the right to object to profiling in specific circumstances, and they must be made aware of this right at the point of data collection.

The General Data Protection Regulation is an attempt to provide protection to us all, as individuals. In an exponentially increasing data driven world we are at risk of being labelled and controlled through the collection, manipulation, and analysis of our data. The law may seem complex and burdensome, but it is the means through which we are granted Rights and freedoms that would not exist otherwise.

As an individual, we may feel powerless to exercise and protect these Rights and freedoms, and this is where Supervisory Authorities, and the judicial system have a role to play. Similarly, Information Systems managers may seek the help and guidance of external bodies to navigate the requirements for compliance. The relevant Supervisory Authorities, as well as organisations such as the European Union Agency for Network and Information Security (ENISA), and the National Cyber Security Centre (NCSC) in the UK, provide best practice advice and frameworks for the protection of data.

Chapter Review

Summary of Key Points

1) Background to data protection.
2) Reasons to protect personal data.
3) 6 Principles of GDPR.
4) Accountability principle.
5) 8 Rights of Natural Persons.

- ❖ What personal information do you share?
- ❖ How would you react if your most intimate secrets were revealed?
- ❖ How would you explain the Principles and Rights to IT?

References

ARTICLE 29 DATA PROTECTION WORKING PARTY. (2017). *Guidelines on Personal data breach notification under Regulation 2016/679.* Directorate General Justice, Brussels.

Commission of the European Union. (2018). *What every business needs to know about the EU's General Data Protection Regulation.* Publications Office of the European Union, Luxembourg.

Cooper, A. (2001). Extraordinary privilege: the trial of Penenden Heath and the Domesday Inquest. *English Historical Review*, 1167–1192.

Council of Europe, E. (1950). Convention for the Protection of Human Rights and Fundamental Freedoms. *Treaty*, 1–26. COE Treaty Office.

Council of the European Union. (2016). Statement of the Council's reasons. *Official Journal of the European Union, 159*(2), 83.

Court of Appeal. (2015). Google Inc v (1) Judith Vidal-Hall (2) Robert Hann (3) Marc Bradshaw. 49. London: Royal Courts of Justice.

European Commission. (1981). Protection of Individuals with regard to Automatic Processing of Personal Data 1981. *Official Journal of the European Union*, 031. Eur-Lex.

European Community Member States. (2000). Charter of Fundamental Rights of the European Union. *Official Journal of the European Communities, 364*, 22.

European Union Member States. (2012). Treaty on the Functioning of the European Union. *Official Journal of the European Union.* Official Journal C 326.

Field, M. (2018, 3 14). Google to ban adverts for cryptocurrency and ICOs. *The Telegraph.*

Griffin, Ricky W; Lopez, Y. (2005). "Bad Behavior" in Organizations: A Review and Typology for Future Research. *Journal of Management, 31*(6), 988–1005.

Hopkins, K. (1980). Taxes and Trade in the Roman Empire (200 B.C.–A.D. 400). *Journal of Roman Studies*(70), 101–125.

Information Commisions Office. (2015, December). *Big Data, Artificial Intelligence, Machine Learning and Data Protection.* Retrieved from ICO.

Mike Butcher. (2017). Cambridge Analytica CEO talks to TechCrunch about Trump, Hillary and the future. *TechCrunch.*

Office for National Statistics. (2018). *Overview of fraud and computer misuse statistics for England and Wales.* Office for National Statistics, London.

Ohlberg, Mareike; Ahmed, Shazeda; Lan, B. (2017). *Central Planning, Local Experiments: The complex implementation of China's Social Credit System.* Mercator Institute for China Studies 12.

Penney, J. (2016). Chilling Effects: Online Surveillance and Wikipedia Use. *Berkeley Technology Law Journal, 31*(1), 117.

Roffe, D. (2000). *Domesday; The Inquest and The Book.* Oxford: Oxford University Press.

Royal Mail Group. (2018). *Marking 500 years of serving the country.* Retrieved from Heritage: https://www.royalmailgroup.com/about-us/heritage

Rubinstein, I. (2013). Big Data: The End of Privacy or a New Beginning? *International Data Privacy Law, 3*(2), 74–87.

Savage, M. (1998). Discipline, Surveillance and the 'Career'. In M. Savage, & K. Mckinlay, Alan & Starkey (Ed.), *Foucault, Management and Organization Theory From Panopticon to Technologies of Self* (1 ed.). London: Sage Publishing.

Sinclair, U. (2004). *The Profits of Religion: An Essay in Economic Interpretation* (Reprint on demand ed.). Whitefish: Kessinger.

Thomson, I. (2018). *When it absolutely, positively needs to be leaked overnight: 120k FedEx customer files spill from AWS S3 silo.* Retrieved from The Register.

Translated by Ingram, G. J. (1912). *The Anglo-Saxon Chronicle.* London: Everyman Press.

Tweed, D. (2018). *Why Governments Might Join the Cryptocurrency Craze*. Retrieved from Bloomberg News: https://www.bloomberg.com/news/articles/2018-02-12/why-governments-might-join-the-cryptocurrency-craze-quicktake

U.S. Federal Trade Commissioner Julie Brill. (2016). Privacy and Data Security in the Age of Big Data and the Internet of Things. *Cyber Security and Privacy Summit* (p. 12). Washington: U.S. Federal Trade Commission.

Vroman, Margaret; Stulz, Karin; Hart, Claudia; Stulz, E. (2016). Employer Liability for Using Social Media in Hiring Decisions. *Journal of Social Media for Organizations, 3*(1).

Appendix

Appendix 1a: GDPR Definitions of Terms:

Article 4

Definitions for the purposes of this Regulation:

(1) **'personal data'** means any information relating to an identified or identifiable natural person ('data subject'); an identifiable natural person is one who can be identified, directly or indirectly, in particular by reference to an identifier such as a name, an identification number, location data, an online identifier or to one or more factors specific to the physical, physiological, genetic, mental, economic, cultural or social identity of that natural person;

(2) **'processing'** means any operation or set of operations which is performed on personal data or on sets of personal data, whether or not by automated means, such as collection, recording, organisation, structuring, storage, adaptation or alteration, retrieval, consultation, use, disclosure by transmission, dissemination or otherwise making available, alignment or combination, restriction, erasure or destruction;

(3) **'restriction of processing'** means the marking of stored personal data with the aim of limiting their processing in the future;

(4) **'profiling'** means any form of automated processing of personal data consisting of the use of personal data to evaluate certain personal aspects relating to a natural person, in particular to analyse or predict aspects concerning that natural person's performance at work, economic situation, health, personal preferences, interests, reliability, behaviour, location or movements;

(5) **'pseudonymisation'** means the processing of personal data in such a manner that the personal data can no longer be attributed to a specific data subject without the use of additional information, provided that such additional information is kept separately and is subject to technical and organisational measures to ensure that the personal data are not attributed to an identified or identifiable natural person;

(6) **'filing system'** means any structured set of personal data which are accessible according to specific criteria, whether centralised, decentralised or dispersed on a functional or geographical basis;

(7) **'controller'** means the natural or legal person, public authority, agency or other body which, alone or jointly with others, determines

the purposes and means of the processing of personal data; where the purposes and means of such processing are determined by Union or Member State law, the controller or the specific criteria for its nomination may be provided for by Union or Member State law;

(8) **'processor'** means a natural or legal person, public authority, agency or other body which processes personal data on behalf of the controller;

(9) **'recipient'** means a natural or legal person, public authority, agency or another body, to which the personal data are disclosed, whether a third party or not. However, public authorities which may receive personal data in the 4.5.2016 EN Official Journal of the European Union L 119/33 framework of a particular inquiry in accordance with Union or Member State law shall not be regarded as recipients; the processing of those data by those public authorities shall be in compliance with the applicable data protection rules according to the purposes of the processing;

(10) **'third party'** means a natural or legal person, public authority, agency or body other than the data subject, controller, processor and persons who, under the direct authority of the controller or processor, are authorised to process personal data;

(11) **'consent'** of the data subject means any freely given, specific, informed and unambiguous indication of the data subject's wishes by which he or she, by a statement or by a clear affirmative action, signifies agreement to the processing of personal data relating to him or her;

(12) **'personal data breach'** means a breach of security leading to the accidental or unlawful destruction, loss, alteration, unauthorised disclosure of, or access to, personal data transmitted, stored or otherwise processed;

(13) **'genetic data'** means personal data relating to the inherited or acquired genetic characteristics of a natural person which give unique information about the physiology or the health of that natural person and which result, in particular, from an analysis of a biological sample from the natural person in question;

(14) **'biometric data'** means personal data resulting from specific technical processing relating to the physical, physiological or behavioural characteristics of a natural person, which allow or confirm the unique identification of that natural person, such as facial images or dactyloscopic data;

(15) **'data concerning health'** means personal data related to the physical or mental health of a natural person, including the provision of health care services, which reveal information about his or her health status;

(16) **'main establishment'** means:
(a) as regards a controller with establishments in more than one Member State, the place of its central administration in the Union, unless the decisions on the purposes and means of the processing of personal data are taken in another establishment of the controller in the Union and the latter establishment has the power to have such decisions implemented, in which case the establishment having taken such decisions is to be considered to be the main establishment;
(b) as regards a processor with establishments in more than one Member State, the place of its central administration in the Union, or, if the processor has no central administration in the Union, the establishment of the processor in the Union where the main processing activities in the context of the activities of an establishment of the processor take place to the extent that the processor is subject to specific obligations under this Regulation;

(17) **'representative'** means a natural or legal person established in the Union who, designated by the controller or processor in writing pursuant to Article 27, represents the controller or processor with regard to their respective obligations under this Regulation;

(18) **'enterprise'** means a natural or legal person engaged in an economic activity, irrespective of its legal form, including partnerships or associations regularly engaged in an economic activity;

(19) **'group of undertakings'** means a controlling undertaking and its controlled undertakings;

(20) **'binding corporate rules'** means personal data protection policies which are adhered to by a controller or processor established on the territory of a Member State for transfers or a set of transfers of personal data to a controller or processor in one or more third countries within a group of undertakings, or group of enterprises engaged in a joint economic activity;

(21) **'supervisory authority'** means an independent public authority which is established by a Member State pursuant to Article 51; L 119/34 EN Official Journal of the European Union 4.5.2016

(22) **'supervisory authority concerned'** means a supervisory authority which is concerned by the processing of personal data because:
(a) the controller or processor is established on the territory of the Member State of that supervisory authority;

(b) data subjects residing in the Member State of that supervisory authority are substantially affected or likely to be substantially affected by the processing; or

(c) a complaint has been lodged with that supervisory authority;

(23) **'cross-border processing'** means either:

(a) processing of personal data which takes place in the context of the activities of establishments in more than one Member State of a controller or processor in the Union where the controller or processor is established in more than one Member State; or

(b) processing of personal data which takes place in the context of the activities of a single establishment of a controller or processor in the Union but which substantially affects or is likely to substantially affect data subjects in more than one Member State.

(24) **'relevant and reasoned objection'** means an objection to a draft decision as to whether there is an infringement of this Regulation, or whether envisaged action in relation to the controller or processor complies with this Regulation, which clearly demonstrates the significance of the risks posed by the draft decision as regards the fundamental rights and freedoms of data subjects and, where applicable, the free flow of personal data within the Union;

(25) **'information society service'** means a service as defined in point (b) of Article 1(1) of Directive (EU) 2015/1535 of the European Parliament and of the Council (1);

(26) **'international organisation'** means an organisation and its subordinate bodies governed by public international law, or any other body which is set up by, or on the basis of, an agreement between two or more countries.

Chapter 2: Organisations, Institutions, and Roles

At a Glance:

- Introduction
    - European Data Protection Supervisor
    - European Data Protection Board: Duties and Hierarchy
- Supervisory Authority: Responsibilities, Powers, and Tasks
    - The Information Commissioner's Office
- Organisations under the GDPR
- The United Kingdom

Learning Objectives: *Students should be able to...*

- Understand the role and functions of key institutions within the EU and nationally.
- Explain the purpose and scope of Supervisory Authorities relative to data protection.
- Describe the role and impact of the various courts and tribunals in the enforcement of data protection legislation.

Key Terms

1) European Data Protection Supervisor
2) European Data Protection Board
3) Supervisory Authorities
4) Public Authorities
5) Non-governmental Organisations
6) Court of Justice of the European Union
7) European Union Agency for Network and Information Security
8) Government Communications Headquarters
9) National Cyber Security Centre
10) Investigatory Powers Commissioner's Office
11) Investigatory Powers Tribunal

Introduction

Quis Custodiet Ipsos Custodes?

"Who watches the watchmen?"

The GDPR requires Member States to establish authorities and bodies whose remit is to facilitate the primary goals of the legislation. The Principles enshrined within the GDPR underpin the goals and duties of the various institutions and supervisory bodies whose role it is to encourage, monitor, and enforce compliance. Understanding the nature and functions of these institutions and bodies is important for the comprehension of the system as a whole. This chapter gives a brief introduction to the supervisory and judicial institutions relevant; at a European Union level and for the UK. The special requirements for public authorities, charities, and non-governmental organisations are also touched upon. An introduction to the role of GCHQ, and its sections, is included for completeness in the understanding of activities undertaken by public bodies. Finally, we look at the role of the newly formed Investigatory Powers Commissioner's Office and the Investigatory Powers Tribunal.

European Union

European Data Protection Supervisor

The European Data Protection Supervisor (EDPS) is the supervisory authority overseeing all European Union institutions and bodies relevant to data protection law. It is responsible for the conduct and compliance of EU public authorities regarding the processing of personal information. Europol, the European body which actively works with law enforcement in Member States and internationally, is also under the supervision of EDPS. As the supervisory authority of the European Union's various organisations, the EDPS sets out to establish and demonstrate best practice for compliance and works with the data protection authorities (DPA's) of each Member State to develop a cohesive approach across the European Union.

The EDPS has three main functions: supervision, consultation, and cooperation.

- ❖ Supervision:
    - o Monitoring the processing of personal data within EU institutions and bodies.
    - o Liaising with Data Protection Officers (DPO's) within those institutions and bodies.
    - o On-site inspections to gauge levels of compliance.
    - o Investigation of complaints raised by individuals, including EU employees.

- ❖ Consultation:
    - o To European Commission, European Parliament, and the Council of the European Union.
    - o Advises on policy areas relevant to data protection issues.
    - o Delivers formal opinion, comments, or policy papers.
    - o Intervenes in cases before the European Court of Justice.

- ❖ Cooperation:
    - o Works closely with the EDPB, previously known as the Article 29 Working Party.
    - o Shares supervision for Eurodac (biometric database of asylum seekers and illegal immigrants) with national data protection authorities.
    - o Works with Working Party on Police and Justice and with related DPA's.

The stated aims of the EDPS are to serve as a centre for excellence, in practice and in law, by enforcing and reinforcing data protection standards laid out by the legislation of the EU. They achieve these aims through a global approach when making recommendations and developing policy guidance. Through instigating, and maintaining, relationships with a wide array of stakeholders nationally and internationally, they work to create an effective culture of data protection based on EU laws.

European Data Protection Board

The European Data Protection Board (EDPB) is an independent body comprised of representatives from the national data protection authorities of each Member State and includes representatives from the EDPS, and the European Commission.

The EDPB is an independent body of the EU which is a legal person. It is represented by its Chair with one member drawn from each of the Supervisory Authorities (SA's) from each of the Member States. Figure 2.1 shows the hierarchical structure of the EDPB and their responsibilities. The Commission of the EU participates in the meetings of the EDPB, however it does not have voting rights. One of its key advisory roles relates to the level of data protection provided in third countries or within international organisations. The EDPB has very similar tasks and powers to SA's.

Duties of the EDPB

The main objectives of the EDPB are to ensure the consistent application of data protection and privacy law across the European Union. This is achieved through the issuance of guidelines for national DPA's and in facilitating effective cooperation between the DPA's of Member States. Where disputes over cross-border processing arise, the EDPB has the power to rule on the issue and its decisions are binding.

Figure 2.1: The Hierarchy of the European Data Protection Board

Composition of the EU Data Protection Board
EU Data Protection Supervisor, Member States (National DP Authority), EU European Commision
Management is elected by Data Protection Board
Chair
Deputy Chair
Responsibility
Ensures Data Protection Law is Applied Consistently Across EU
Ensures Cooperation Amongst Data Protection Authorities
Authority
Binding Decisions, Professional Opinions, and General Guidelines

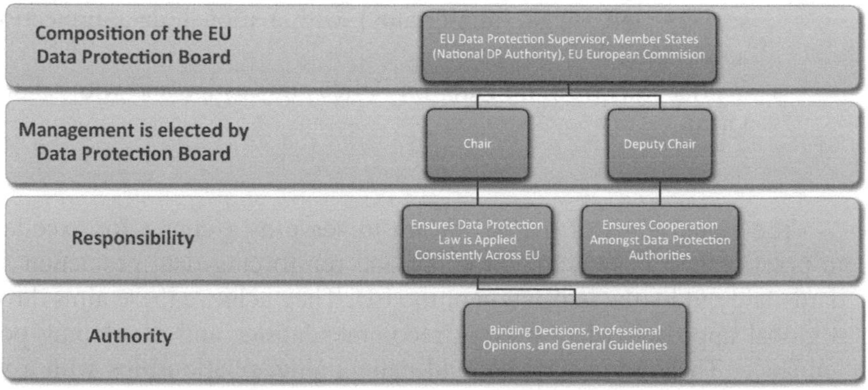

Binding Decisions: Mostly used to settle cross-border disputes over data protection, such as cross-border processing of data. The binding decision mechanism was introduced into the framework by the GDPR.

Professional Opinion: The EDPB publishes its opinion regarding industry Codes of Conduct on compliance with data protection regulations.

General Guidelines: The interpretation of the law, where there is need for clarification. The relevant authorities cooperate to develop guidelines, which are then approved by the EDPB by majority vote.

Supervisory Authorities

Supervisory Authorities (SA) are institutions or organisations established in the public interest, by governments, to uphold the principles and rights within specific legislative areas. The GDPR speaks directly to the role of supervisory bodies in the upholding and maintenance of its requirements. These National Data Protection Authorities are primarily responsible for the protection of information privacy in their respective countries through various activities designed to educate, monitor, and enforce the law.

Each Member State shall provide for one or more independent public authorities to be responsible for monitoring the application of this regulation, in order to protect the fundamental rights and freedoms of natural persons in relation to processing and to facilitate the free flow of personal data within the Union

Article 5 (1)

Depending on the specific constitutional requirements of each Member State, there may be the need to establish multiple SA's. Where this is the case a single lead authority must be named, and this lead authority will represent the Member State on the EDPB. Independence from any external influence is of paramount importance in the lawful functioning of the SA. This requirement ensures the ability of the SA to fulfil its duties without prejudice or bias to the entity being investigated; whether governmental, commercial, or otherwise.

The independent Supervisory Authority is an essential element for the application and protection of the fundamental Rights and freedoms of natural persons under the GDPR. This is reflected in Article 54, which calls for each Member State to establish laws for the creation and functioning of the SA. They must cooperate with one another to contribute to the further development of data protection within the EU and to ensure a consistent approach. This cooperation should be achieved without the need for formal agreements between Member States; to this end the SA's in each Member State will have similar tasks and powers.

The need for independence is supported in the transparent manner in which members of the SA are appointed. Whilst they must have relevant experience working in data protection, they are not allowed to work (paid or otherwise) in any occupation which compromises their duties within the SA. To protect members of the SA from external pressures, they may be dismissed only where they have exhibited gross misconduct, or where they can no longer fulfil their duties.

'Supervisory authority concerned' means a supervisory authority which is concerned by the processing of personal data because:

a) The controller or processor is established on the territory of the member state of that supervisory authority
b) Data subjects residing in the member state of that supervisory authority are substantially affected of likely to be substantially affected by the processing; or
c) A complaint has been lodged with that supervisory authority

Article 4 (22)

Duties and tasks of the Supervisory Authority

Some of the major tasks for the SA include:

- To inform, relevant to data processing; the public, individuals, controllers and processors, of the: risks, rules, safeguards, and rights through:
  - Promoting awareness of data protection issues.
  - Monitoring and enforcing the regulation through conducting investigations, which arise from complaints or other sources.
  - The creation of Codes of Practise and the development of approved certification.

Powers of the Supervisory Authorities

The powers awarded to Supervisory Authorities under the GDPR are greater and wider reaching than under previous data protection legislation. They have the power to issue warnings and reprimands. They can also order data controllers to:

- Come into compliance with the regulation.
- Impose a limitation on processing.
- Suspend data flows to third parties.

Relevant certification can be withdrawn by the SA and an administrative fine may also be imposed. The maximum fine allowed is equivalent to 4% of the global turnover of the offending organisation. Further action, through the instigation of legal proceedings against a Controller, or the involvement of the judicial authorities are also within the powers of the Supervisory Authorities.

Information Commissioner's Office

> *"Those who self-report, who engage with the ICO to resolve issues, and who can demonstrate effective accountability arrangements, can expect this to be taken into account when we consider any regulatory action."*
>
> Elizabeth Denham, UK Information Commissioner

In the United Kingdom the relevant supervisory authority for national data protection is the Information Commissioner's Office (ICO). The ICO is an independent public body funded by government through the Department for Digital, Culture, Media, and Sport (DCMS). The ICO's mission is grounded in the protection of information rights and data privacy, whilst encouraging openness by public bodies.

The duties of the ICO are many and varied and can be broadly ranged as the following:

- Maintain a register of data controllers.
- Upholding and reinforcing relevant legisl ation.
- Handling concerns and complaints.
- Acting on data protection law.
- Privacy and Electronic Communications regulations.
- Freedom of Information and Environmental information.
- International work.
- Maintaining a grants program.

The ICO in Action

Charities breaking the law

Even charities knowingly break data protection law. The ICO investigated the activities of British charities when collecting and processing the personal data of their financial donors. Over the period between 2015 to 2017, a total of thirteen charities were found breaking the law; primarily in the areas of fundraising and nuisance phone calls. Of the charities breaking the law, eleven were fined by the ICO.

Organisations Under the GDPR

Many organisations are affected under the new data protection legislation this section will briefly describe a few of the organisations under a few broad subheadings (specifically mentioned within the regulation).

Public Authorities

The GDPR does not specify what defines a public authority, and thus it will be left to each Member State (and complying nations) to define. This ambiguity is problematic in that the definition of "public authority" will differ across

Europe leading to contradictions and confusion, especially in the choice of the lawfulness of processing. Public authorities are not permitted to use the "legitimate interest" purpose, and the ICO has advised that the "consent" grounds for processing will not be appropriate due to the imbalance of power between individuals and state or public authorities. Furthermore, public authorities must appoint a Data Protection Officer (DPO), and this entails specific requirements and obligations. The requirements, role, and functions of the DPO are detailed in Chapter 7.

Types of Public Authorities

For the purposes of the GDPR public authorities include:

- The Armed forces.
- State government departments.
- Legislative bodies.
- Local governments.
- Police forces.
- The National Health Service.
- Institutions of Higher Education.
- Maintained schools.
- Publicly owned companies.
- "Other" public bodies.

NGO's and Charities

Charities and Non-Government organisations share a concern for social issues. They are usually run as not-for-profit organisations, but not in every case. Whilst there may be a self-professed desire to "do good" within society, these organisations are subject to the same data protection laws as all other organisations. These organisations may work locally, nationally, or globally and the very nature of the organisation's sphere of activity may entail stricter controls over the way personal data is handled, for example those organisations dealing with ex-offenders.

A charity is an organisation whose primary objective is the amelioration of a social need through activities arising from the desire to help humanity. This type of activity is considered as a common good; that is in the interest of society at large.

Charities are non-profit organisations (NPO) and are required to adhere to the specific laws regarding their area of activity. Some of the areas in which charities work are:

- Education
- Health
- Environment
- Spirituality
- Culture

Similar to Charities, non-government organisations (NGO's), operate independently from the government, but work to further specific social or political agendas. These social and political spheres of activity often overlap areas of governmental activity and interest. When run under a not-for-profit model, funds are raised through donations, grants, and any other activities which can help fund their work.

In the UK, the ICO has reviewed the guidance on fundraising for Charities and NGO's which was drawn up by the Institute of Fundraising (IoF) and the Fundraising Regulator. This guidance provides valuable information for these types of organisations in the compliance to the GDPR.

NGO's and Charities as Data Controllers

Charities and NGO's process personal data in a number of ways, some of which are:

- Processing the personal data of employees.
- Processing the personal data of trustees.
- Processing the perso nal data of volunteers.
- Processing the personal data of donors and supporters.
- As a provider of personalised services to beneficiaries.
- As a provider of personalised services clients.

Where individuals engage in fundraising for an organisation, and personal data is handled in aid of that activity, that individual is also considered a data controller under the GDPR. This remains the case even when the individual is acting independently of the charity or NGO for whom the funds are being raised.

Institutions and Agencies

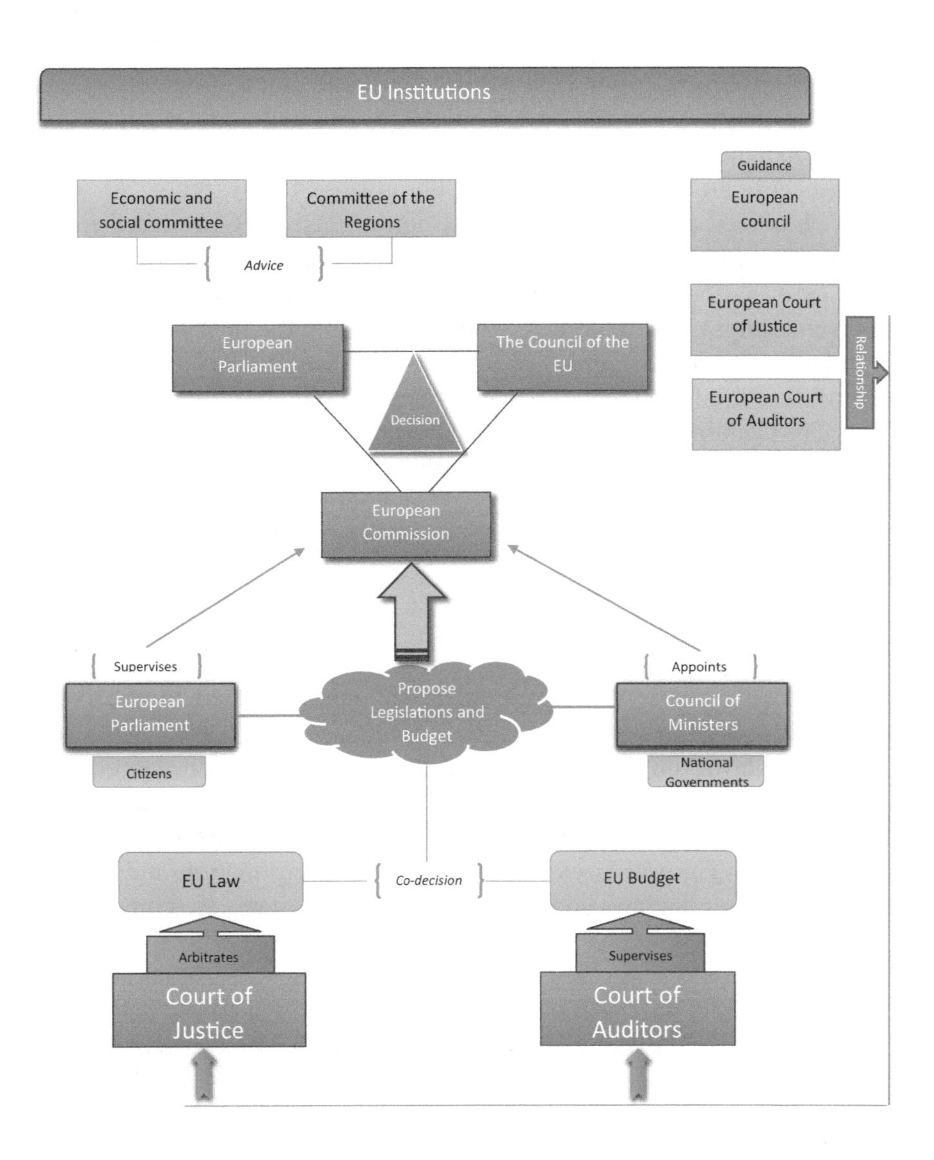

Court of Justice of the European Union

The Court of Justice of the European Union (CJEU) interprets the laws of the EU to ensure its consistent application throughout the Member States of the Union. It is comprised of 28 judges, one from each Member State, and 11 advocates general. Most often, it is the court of last resort in legal disputes between the EU and national governments. However, in specific cases, individuals, organisations, or companies, can apply to the court where they believe an institution of the EU has infringed their rights.

The CJEU is comprised of the Court of Justice (CoJ) and the General Court. The Court of Justice is used by the national courts of Member States, and other nations, when preliminary rulings are requested. Preliminary rulings are sought where there is doubt surrounding the interpretation of a particular EU law and a national court seeks clarification and guidance. National courts also apply to the CoJ for annulments and in appeals

The General Court deals primarily with cases involving State aid, trade, agriculture, competition law, and trademarks; as this is the court to which actions by individuals, organisations, and companies are brought. In some cases, the governments of the EU may also bring actions to the General Court.

Duties and tasks of the CJEU

---

1) *Interpret the law:* Preliminary rulings, requested by the courts of EU Member States, are attempts to clarify the correct application of EU law. Since EU countries are required to uphold the laws of the EU, they must ensure its proper application. However, the diverse courts within all the nations of the EU may, and do, interpret the law differently. The CJEU clarifies the correct interpretation and determines whether the laws and practices of Member States are compatible with EU law.
2) *Enforcing the law:* Infringement proceedings are actions taken against a national govrnment where they have not complied with EU law. Cases are brought by the European Commission, or by the governments of EU Member States. Where a Member State is found to be in breach of EU law, they must rectify the situation immediately, or face further action before the court and a fine.
3) *Annulling acts:* Actions for annulment may be raised where a legal act of the EU is believed to be in violation of fundamental rights, or EU treaties. The government of a Member State, the Council of the

European Union, the European Commission, and even the European Parliament may ask the court to annul an act. Where an act directly concerns a private individual, they may also apply to court for annulment.

4) *Ensuring action:* Governments and institutions of the EU can complain to the court where the Parliament of the EU, the Council of the EU, or the European Commission fail to take action under circumstances where they are obliged to do so. It is also possible for private individuals and organisations to complain to the court, in limited cases.
5) *Sanctioning institutions:* Actions for damages may be brought before the court by any private individual or organisation whose interest has been harmed by the EU. The sanction may be based on an EU institution's action, or failure to act, or the same by staff of that institution. Actions for damages are claims for financial recompense.

Example: When is an IP address personal data?

In October 2016, the CJEU declared that dynamic IP addresses are to be considered as personal data when registered by the operator of a website. The court noted that the IP address plus extra information from the Internet Service Provider (ISP) of the user could be used to identify the user.

This judgement was the final instalment of a saga which began in 2008, when a German citizen; Patrick Breyer, asked the German courts to stop government operated websites from storing his IP address. The German Government; the *Bundesrepublik Deutschland*, argued that this type of activity was important in the prevention of cyber-attacks, and in the prosecution of those involved in internet piracy. After making its way through the German judicial system, Mr. Breyer was vindicated in his belief that this activity, by the website operators, was a breach of data protection laws.

European Union Agency for Network and Information Security: ENISA

The European Union Agency for Network and Information Security (ENISA) is an EU centre of expertise in the field of Cyber Security. Helping the nations and institutions of the EU to prevent, detect, and respond, to problems relating to information security. One of ENISA's main goals is to assist the functioning of the EU's internal market through the development, at a high level, of network and information security (NIS) awareness.

Working with both Member States and private sector organisations, ENISA provides advice and practical solutions through the development of National Cyber Security strategies, and in the area of EU policy implementation for NIS. Findings on cyber security issues are regularly published in the form of studies and reports.

ENISA's activities are grouped into three main areas:

- Recommendations.
- Supporting policy making and implementation.
- Collaboration with operational teams throughout the EU.

ENISA is involved in a wide range of activities relevant to its remit of raising awareness of NIS issues and engendering a culture of network and information security. These activities are:

- CSIRTs: Working in collaboration with CSIRTs across Europe.
- Cyber Crisis Management: Pioneering and developing consistent mechanisms.
- Cyber Security Education: Promoting NIS awareness through knowledge.
- Incident Reporting: Based on the following regulations;
  - Art. 13a Telecom Framework Directive
  - Art. 19 eIDAS regulation
  - Art. 16 (4) NIS Directive
- Standards and Certification: Assists the development of ICT security frameworks.
- Threat and Risk Management: Overview of current and emerging threats and risks.
- Training Cyber Security Specialists: Provision of training materials.

The United Kingdom

Government Communications Headquarters

The Government Communications Headquarters (GCHQ) is the UK intelligence and security body responsible for the provision of Signals Intelligence (SIGINT), and information assurance, to the UK government and its armed forces. It has a twofold mission; to gather intelligence and to protect the communications of the UK. It is under the remit of the Secretary of State for Foreign and Commonwealth Affairs of the UK. GCHQ is composed of two parts; the National Cyber Security Centre (NCSC), and the Composite Signals Organisation (CSO), now known as GCHQ Bude.

Within the Government, overall responsibility for intelligence and security issues (and their respective agencies) falls to the Prime Minister. The minister responsible for the everyday running of the GCHQ, is the Foreign Secretary. Funding for the three security and intelligence agencies; GCHQ, MI5, and Secret Intelligence Service (MI6), is provided through the Single Intelligence Account, and the National Cyber Security Programme.

The Intelligence Services Act 1994

Part 3) The Government Communications Headquarters:

(1) *There shall continue to be a Government Communications Headquarters under the authority of the Secretary of State; and, subject to subsection (2) below, its functions shall be—*

(a) *to monitor [make use of] or interfere with electromagnetic, acoustic* and other *emissions and any equipment producing such emissions and to obtain and provide information derived from or related to such emissions or equipment and from encrypted material; and*

(b) *to provide advice and assistance about:*

(i) *languages, including terminology used for technical matters, and*

(ii) *cryptography and other matters relating to the protection of information and other material, to the armed forces of the Crown, to Her Majesty's Government in the United Kingdom or to a Northern Ireland Department [or, in such cases as it considers appropriate, to other organisations or persons, or to the general public, in the United Kingdom or elsewhere.]*

(2) *The functions referred to in subsection (1)(a) above shall be exercisable only—*

(a) *in the interests of national security, with particular reference to the defence and foreign policies of Her Majesty's Government in the United Kingdom; or*

(b) *in the interests of the economic well-being of the United Kingdom in relation to the actions or intentions of persons outside the British Islands; or*

(c) *in support of the prevention or detection of serious crime.*

(3) In this Act the expression "GCHQ" refers to the Government Communications Headquarters and to any unit or part of a unit of the armed forces of the Crown which is for the time being required by the Secretary of State to assist the Government Communications Headquarters in carrying out its functions.

Many of GCHQ's activities involve collaborating with other organisations. Primarily this involves working closely with the other security and intelligence agencies within the UK (MI5 and MI6), and with similar agencies from other nations. Sharing knowledge and expertise is an important factor in countering the increasingly complex threats to national security which arise locally and globally. Working together with other Intelligence agencies may seem an obvious activity for GCHQ to engage in, however they also work in other, less obvious areas, such as:

Partnering with academia:

- To help overcome rising cyber security challenges.
- To draw on current academic research.
- In developing courses to increase the number of suitably trained staff.

Supporting education in cyber security:

- Partnerships with the schools around the Cheltenham area.
- Working with educational bodies across the UK.

Local community partnerships:

- Voluntary projects.
- Neighbourhood regeneration.
- Supporting local, national, international charities.

GCHQ has six sites across the United Kingdom, the most famous of which is the "Doughnut" in Cheltenham, England, where the majority of GCHQ's staff are based. GCHQ Bude, based in Cornwall, and GCHQ Scarborough, based in North Yorkshire, provide signals intelligence. RAF Menwith Hill, based in Harrogate, North Yorkshire, provides intelligence to the UK, the USA, and their allies. The National Cyber Security Centre is based in London and provides protection from cyber-attacks on the UK's critical services and infrastructure. GCHQ Manchester, based in Greater Manchester, provides "real-time" response capabilities to emerging and active cyber threats.

The National Cyber Security Centre

The National Cyber Security Centre (NCSC) was established in 2016 to provide a single point of contact for advice and support to both public and private sectors on issues pertinent to Cyber Security. As a branch of GCHQ, it brings the information assurance arm of GCHQ, known by the acronym CESG, together with the Centre for Cyber Assessment, the CERT-UK, and the Centre for Protection of National Infrastructure. Its main objective is to protect critical services against the threat of cyber-attacks and to respond to any such attacks.

The NCSC achieves this in part by:

- Managing major incidents.
- Technological improvements to UK internet security.
- Advising citizens and organisations.

In responding to cyber security incidents, the NCSC's focus is on harm reduction; to the UK in general, and to organisations specifically. A risk-based approach to cyber security highlights the need for active risk reduction by the NCSC.

Risk reduction is engendered through:

- Securing Public and Private Networks.
- Provision of Practical Guidance to all.
- Ongoing engagement with industry and academic experts.

The NCSC aims to protect information and systems vital to the smooth functioning of the United Kingdom. It publishes guidance on cyber security issues for use by; government, agencies, critical national infrastructure, and their respective supply chains.

The GCHQ Bude: GCHQ Composite Signals Organisation Morwenstow

Formerly known as the Composite Signals Organisation (CSO), the main listening station for the UK's security services is GCHQ Bude. It gathers information for the detection of threats to the UK; threats arising from within and without. Activities are focussed on the areas of serious crime, terrorism, and espionage, and to that end they work closely with the National Crime Agency (NCA) and Her Majesty's Revenue and Customs department (HMRC).

Work with other government departments covers a broad array of issues:

- Cybercrime.
- Sexual exploitation of children.
- Organised immigration crime.
- Money laundering.
- Financial crime.
- Drugs smuggling.

GCHQ Bude advises government departments on the protection of networks, based on information gathered as part of their ongoing threat detection and monitoring activities.

Structure of the UK Intelligence Services

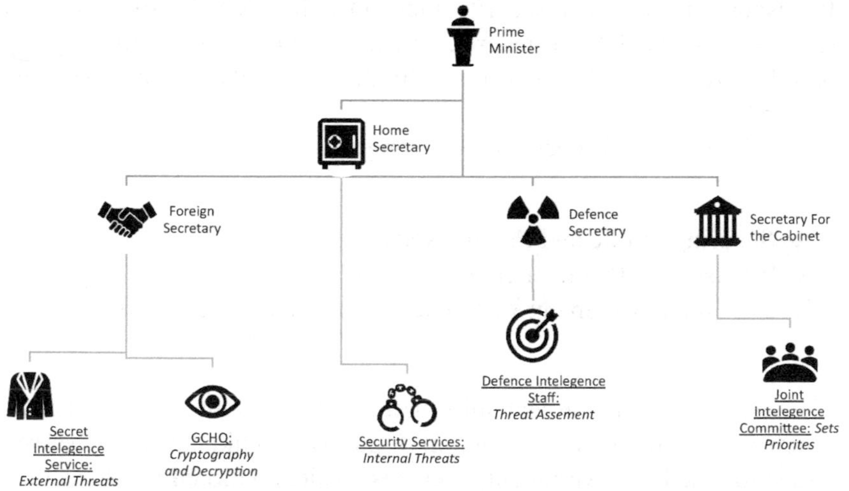

Prime Minister
Home Secretary
Foreign Secretary
Defence Secretary
Secretary For the Cabinet
Secret Intelegence Service: *External Threats*
GCHQ: *Cryptography and Decryption*
Security Services: *Internal Threats*
Defence Intelegence Staff: *Threat Assement*
Joint Intelegence Committee: *Sets Priorites*

Investigatory Powers Commissioner's Office

> *"From today, and for the first time, investigatory powers will be overseen by a single body applying a consistent, rigorous and independent inspection regime across public authorities. This is an important milestone as we start to implement the new oversight powers set out in the Investigatory Powers Act."*
>
> Lord Justice Fulford, Investigatory Powers Commissioner, 2017

Established under the terms of the *Investigatory Powers Act 2016*, the Investigatory Powers Commissioners Office (IPCO) was created to support the work of the Investigatory Powers Commissioner and the Judicial Commissioners. These commissioners oversee the use, by over 600 public authorities and institutions, of investigatory powers, through the monitoring of records, inspections, and monetary penalties. There are three main areas of oversight; communication data, intelligence, and surveillance.

This responsibility for oversight of public authorities includes:

- Law enforcement.
- The security and intelligence agencies.
- The National Crime Agency.
- HM Revenue and Customs.
- Prisons.
- Local authorities.
- Regulators.
- Other government agencies.

In September 2017 the IPCO replaced the previous three oversight bodies; the Interception of Communications Commissioner's Office (IOCCO), the Intelligence Services Commissioner (ISC), and the Office of Surveillance Commissioners (OSC). The IPCO reports to the Prime Minister and is funded by the Home Office and although it is supported by the government, it must remain rigorously independent of any external political influence.

One of the functions of the Judicial Commissioners is the granting of prior approval to public authorities and institutions for activities related to:

- Bulk acquisition of communications data.
- Bulk personal datasets.
- Communications data retention notices.
- Equipment interference.
- Interception.
- National security notices.
- Technical capability notices.

The Technology Advisory Panel assists the Judicial Commissioners in the assessment of various technical issues raised during a request for prior approval of investigatory activities. It is made up of legal, technical, and scientific experts from various fields.

Investigatory Powers Tribunal

The Investigatory Powers Tribunal (IPT) is an independent judicial body that is a court and thus appointments to the IPT are essentially judicial in nature. These appointments vary depending on whether the proposed candidate is a serving member of the senior judiciary of England and Wales, Scotland or Northern Ireland, referred to as a "judicial member", or if they are a "non-judicial member". A non-judicial member could be, a former member of the judiciary who is no longer serving or, a senior member of the legal profession who is not a full-time judge. The IPT values the diversity of experience and expertise that having both judicial and non-judicial members brings to the Tribunal.

An important piece of legislation is the *Regulation of Investigatory Powers Act* 2000 (RIPA). It is the main source of law that establishes and regulates the power of public bodies to intrude upon the privacy of members of the public. RIPA provides for the establishment of a body to which citizens may complain when these powers are believed to have been used unlawfully: this is the function of the Investigatory Powers Tribunal.

Section 65 of RIPA states that the Investigatory Powers Tribunal (IPT) has judicial oversight over the activities and conduct of the UK's intelligence services. Its jurisdiction covers the Secret Intelligence Service (MI6), the Security Service (MI5), and the GCHQ, and is the only independent body to whom complaints about the activities of the security intelligence services may be made. It is this court which investigates complaints about unlawful covert techniques used by public authorities and conduct which breaches human rights law.

The IPT is unique as a court of its kind in that it operates an 'inter partes' system for hearings. This allows *both* parties to present their arguments on the assumption that the facts alleged in the complaint are true. Hearings of this kind allow the IPT to consider the important legal issues around the case before the facts have been proved. The court's ability to consider evidence that would be otherwise inadmissible in any other court, gives it an investigatory role. This is part of an *inquisitorial* process, which allows the court to ascertain the facts of the complaint; in stark contrast to the *adversarial* process used in ordinary court proceedings. The inquisitorial process is enabled, in part, due to the sensitive nature of the material involved.

The IPT has a duty to protect any sensitive material it receives from the Security Intelligence Agencies (SIA). This duty gives it the power to review and obtain sensitive evidence with greater freedom than is available to the ordinary courts. At times, the complainant may not be permitted to hear or see the evidence obtained by the IPT, or even be made aware of the nature of that evidence. Likewise, any documents the complainant provides to the IPT will not be disclosed without consent.

Encouraging members of the public to bring complaints before the court is a primary concern to the IPT, and to this end the provision of strict confidentiality and identity protection for the complainant, and witnesses, is assured. The use of the court if free of charge to the complainant, and there is no requirement to engage a lawyer; although the complainant is free to do so. Where the case is lost by the complainant, the court has never awarded costs to the public authority. This reinforces the IPT's desire to encourage people to bring a complaint before the court.

The IPT has wide ranging powers regarding remedy for the complainant. It may order a stop to the activity complained of, or for material to be destroyed. It is also within its powers to grant compensation to the complainant. The only higher route for appeal against a decision by the IPT, is to the European Court of Human Rights, in Strasbourg.

A recent case: *Dias and others V Cleveland Police*

In 2017 the IPT found that the Cleveland Police force, based in North Yorkshire, behaved unlawfully when they used surveillance powers to seize the phone records of two former police officers. The ex-officers; Mark Dias and Stephen Matthews, were investigated by Cleveland Police in an attempt to identify "whistle blowers" who had leaked information regarding institutional racism within the force. The Tribunal found that there was no legal basis for the seizing of phone records and awarded the ex-officers and other complainants £3000 each.

Chapter Review

Summary of Key Points

1) Role of key European institutions.
2) Functions of Supervisory Authorities.
3) Requirements for Public Authorities.
4) Authority of the CJEU.
5) Role of ENISA.
6) Outline of GCHQ and NCSC.
7) Role of the IPC Office.
8) Scope of the IP Tribunal.

- Which institution's decisions are binding?
- Think about the interaction between the ICO and the EDPB.
- Who is the SA in your country? Is there only one?
- Do you work for a PA, NGO or neither?

References

ARTICLE 29 DATA PROTECTION WORKING PARTY. (2016). *Guidelines for identifying a controller or processor's lead supervisory authority.* Directorate General justice and Consumers, Brussels.

Bowcott, O. (2017, 7 1). UK surveillance and spying watchdog begins work. *The Guardian*.

Court of Justice of the European Union. (2016). Patrick Breyer v Bundesrepublik Deutschland. 2. Luxembourg: InfoCuria.

Court of Justice of the European Union. (2018). *General Presentation.* Retrieved from The Institution: https://curia.europa.eu/jcms/jcms/Jo2_6999/en/

Denham, E. (2018). Information rights and responsibilities, with the Information Commissioner. *Association of Chief Executives and Public Chairs' Forum* . London: Information Commissioner's Office.

European Commission. (2018). *What is the European Data Protection Board (EDPB)?* Retrieved from Website: https://ec.europa.eu/info/law/law-topic/data-protection/reform/rules-business-and-organisations/

enforcement-and-sanctions/enforcement/what-european-data-protection-board-edpb_en

European Data Protection Supervisor. (2018). *Data Protection*. Retrieved from Website: https://edps.europa.eu/data-protection_en

European Union. (2018). *Court of Justice of the European Union (CJEU)*. Retrieved from About the EU: Institutions and Bodies: https://europa.eu/european-union/about-eu/institutions-bodies/court-justice_en

GCHQ. (2018). *Who we are*. Retrieved from Website: https://www.gchq.gov.uk/who-we-are

Information Commissioner's Office. (2018). *Charity fundraising enforcement action*. Retrieved from Website: https://ico.org.uk/action-weve-taken/charity-fundraising-enforcement-action/

Investigatory Powers Commissioner's Office. (2018). *Who we are*. Retrieved from Website: https://www.ipco.org.uk/

Investigatory Powers Tribunal. (2017). DIAS AND OTHERS V CLEVELAND POLICE. London: The Investigatory Powers Tribunal.

Investigatory Powers Tribunal. (2018). *General Overview and Background*. Retrieved from Website: http://www.ipt-uk.com/

National Cyber Security Centre. (2018). *What we do*. Retrieved from Website: https://www.ncsc.gov.uk/

Roland, N. (2016, 10). GDPR: The Data Protection Supervisor(s): Who are you? Where are you? *Lexology*.

Sosin, J. (2000). Ausonius' Juvenal and the Winstedt fragment. *Classical Philology, 95*(2), 199–206.

The European Union Agency for Network and Information Security. (2018). *About ENISA*. Retrieved from Website: https://www.enisa.europa.eu/about-enisa

The Information Commissioner's Office (ICO). (2018). *Guide to the General Data Protection Regulation (GDPR)*. Retrieved from Website: https://ico.org.uk/for-organisations/guide-to-the-general-data-protection-regulation-gdpr/

UK Government. (2016). *NATIONAL CYBER SECURITY STRATEGY 2016–2021*. HM Government, London.

United Kingdom. (1994). Intelligence Services Act 1994. legislation.gov.uk.

United Kingdom. (2000). Regulation of Investigatory Powers Act 2000 (RIPA. legislation.gov.uk.

United Kingdom. (2016). Investigatory Powers Act 2016. *legislation.gov.uk*.

Zuiderveen Borgesius, F. (2017). Breyer Case of the Court of Justice of the European Union: IP Addresses and the Personal Data Definition (Case Note). *European Data Protection Law Review, 3*(1), 1.

Appendix

Appendix 2a: List of Supervisory Authorities in Europe

Austria
Österreichische Datenschutzbehörde
Hohenstaufengasse 3
1010 Wien
Tel. +43 1 531 15 202525
Fax +43 1 531 15 202690
e-mail: dsb@dsb.gv.at
Website: http://www.dsb.gv.at/
Dr Andrea JELINEK, Director, Österreichische Datenschutzbehörde

Belgium
Commission de la protection de la vie privée
Rue de la Presse 35
1000 Bruxelles
Tel. +32 2 274 48 00
Fax +32 2 274 48 10
e-mail: commission@privacycommission.be
Website: http://www.privacycommission.be/

Bulgaria
Commission for Personal Data Protection
2, Prof. Tsvetan Lazarov blvd.
Sofia 1592
Tel. +359 2 915 3523
Fax +359 2 915 3525
e-mail: kzld@cpdp.bg
Website: http://www.cpdp.bg/
Mr Ventsislav KARADJOV, Chairman of the Commission for Personal Data Protection
Ms Mariya MATEVA

Croatia
Croatian Personal Data Protection Agency
Martićeva 14
10000 Zagreb
Tel. +385 1 4609 000
Fax +385 1 4609 099

e-mail: azop@azop.hr or info@azop.hr
Website: http://www.azop.hr/
Mr Anto RAJKOVAČA, Director of the Croatian Data Protection Agency

Cyprus
Commissioner for Personal Data Protection
1 Iasonos Street,
1082 Nicosia
P.O. Box 23378, CY-1682 Nicosia
Tel. +357 22 818 456
Fax +357 22 304 565
e-mail: commissioner@dataprotection.gov.cy
Website: http://www.dataprotection.gov.cy/
Ms Irene LOIZIDOU NIKOLAIDOU
Mr Constantinos GEORGIADES

Czech Republic
The Office for Personal Data Protection
Urad pro ochranu osobnich udaju
Pplk. Sochora 27
170 00 Prague 7
Tel. +420 234 665 111
Fax +420 234 665 444
e-mail: posta@uoou.cz
Website: http://www.uoou.cz/
Ms Ivana JANŮ, President of the Office for Personal Data Protection
Mr Ivan PROCHÁZKA, Adviser to the President of the Office

Denmark
Datatilsynet
Borgergade 28, 5
1300 Copenhagen K
Tel. +45 33 1932 00
Fax +45 33 19 32 18
e-mail: dt@datatilsynet.dk
Website: http://www.datatilsynet.dk/
Ms Cristina Angela GULISANO, Director, Danish Data Protection Agency (Datatilsynet)
Mr Christian Vinter HAGSTRØM, Head of Section

Estonia
Estonian Data Protection Inspectorate (Andmekaitse Inspektsioon)
Väike-Ameerika 19
10129 Tallinn
Tel. +372 6274 135
Fax +372 6274 137
e-mail: info@aki.ee
Website: http://www.aki.ee/en
Mr Viljar PEEP, Director General, Estonian Data Protection Inspectorate
Ms Kaja PUUSEPP

Finland
Office of the Data Protection Ombudsman
P.O. Box 315
FIN-00181 Helsinki
Tel. +358 10 3666 700
Fax +358 10 3666 735
e-mail: tietosuoja@om.fi
Website: http://www.tietosuoja.fi/en/
Mr Reijo AARNIO, Ombudsman of the Finnish Data Protection Authority
Ms Elisa KUMPULA, Head of Department

France
Commission Nationale de l'Informatique et des Libertés - **CNIL**
8 rue Vivienne, CS 30223
F-75002 Paris, Cedex 02
Tel. +33 1 53 73 22 22
Fax +33 1 53 73 22 00
Website: http://www.cnil.fr/
Ms Isabelle FALQUE-PIERROTIN, President of CNIL
Ms Florence RAYNAL

Germany
Die Bundesbeauftragte für den Datenschutz und die Informationsfreiheit
Husarenstraße 30
53117 Bonn
Tel. +49 228 997799 0; +49 228 81995 0
Fax +49 228 997799 550; +49 228 81995 550
e-mail: poststelle@bfdi.bund.de
Website: http://www.bfdi.bund.de/

The competence for complaints is split among different data protection supervisory authorities in Germany. Competent authorities can be identified according to the list provided under
https://www.bfdi.bund.de/bfdi_wiki/index.php/
Aufsichtsbeh%C3%B6rden_und_Landesdatenschutzbeauftragte
Ms Andrea VOSSHOFF, Federal Commissioner for Freedom of Information
Prof. Dr. Johannes CASPAR, representative of the federal states.

Greece
Hellenic Data Protection Authority
Kifisias Av. 1-3, PC 11523
Ampelokipi Athens
Tel. +30 210 6475 600
Fax +30 210 6475 628
e-mail: contact@dpa.gr
Website: http://www.dpa.gr/
Mr Petros CHRISTOFOROS, President of the Hellenic Data Protection Authority
Dr. Vasilios ZORKADIS, Director

Hungary
Data Protection Commissioner of Hungary
Szilágyi Erzsébet fasor 22/C
H-1125 Budapest
Tel. +36 1 3911 400
e-mail: peterfalvi.attila@naih.hu
Website: http://www.naih.hu/
Art 29 WP Member: Dr Attila PÉTERFALVI, President of the National Authority for Data Protection and Freedom of Information.
Mr Endre Győző SZABÓ, Vice-president of the National Authority for Data Protection and Freedom of Information.

Ireland
Data Protection Commissioner
Canal House
Station Road
Portarlington
Co. Laois
Lo-Call: 1890 25 22 31
Tel. +353 57 868 4800

Fax +353 57 868 4757
e-mail: info@dataprotection.ie
Website: http://www.dataprotection.ie/
Ms Helen DIXON, Data Protection Commissioner
Mr John O'DWYER, Deputy Commissioner; Mr Dale SUNDERLAND, Deputy Commissioner

Italy
Garante per la protezione dei dati personali
Piazza di Monte Citorio, 121
00186 Roma
Tel. +39 06 69677 1
Fax +39 06 69677 785
e-mail: garante@garanteprivacy.it
Website: http://www.garanteprivacy.it/
Mr Antonello SORO, President of Garante per la protezione dei dati personali Ms Vanna PALUMBO, Head of Service for EU and International Matters

Latvia
Data State Inspectorate
Director: Ms Signe Plumina
Blaumana str. 11/13–15
1011 Riga
Tel. +371 6722 3131
Fax +371 6722 3556
e-mail: info@dvi.gov.lv
Website: http://www.dvi.gov.lv/
Ms Signe PLUMINA, Director of Data State Inspectorate
Ms Aiga BALODE

Lithuania
State Data Protection
Žygimantų str. 11–6a
011042 Vilnius
Tel. + 370 5 279 14 45
Fax +370 5 261 94 94
e-mail: ada@ada.lt
Website: http://www.ada.lt/
Mr Algirdas KUNČINAS, Director of the State Data Protection Inspectorate

Ms Neringa KAKTAVIČIŪTĖ-MICKIENĖ, Head of Complaints Investigation and International Cooperation Division

Luxembourg
Commission Nationale pour la Protection des Données
1, avenue du Rock'n'Roll
L-4361 Esch-sur-Alzette
Tel. +352 2610 60 1
Fax +352 2610 60 29
e-mail: info@cnpd.lu
Website: http://www.cnpd.lu/
Ms Tine A. LARSEN, President of the Commission Nationale pour la Protection des Données
Mr Thierry LALLEMANG, Commissioner

Malta
Office of the Data Protection Commissioner
Data Protection Commissioner: Mr Joseph Ebejer
2, Airways House
High Street, Sliema SLM 1549
Tel. +356 2328 7100
Fax +356 2328 7198
e-mail: commissioner.dataprotection@gov.mt
Website: http://www.dataprotection.gov.mt/
Mr Saviour CACHIA, Information and Data Protection Commissioner
Mr Ian DEGUARA, Director – Operations and Programme Implementation

Netherlands
Autoriteit Persoonsgegevens
Prins Clauslaan 60
P.O. Box 93374
2509 AJ Den Haag/The Hague
Tel. +31 70 888 8500
Fax +31 70 888 8501
e-mail: info@autoriteitpersoonsgegevens.nl
Website: https://autoriteitpersoonsgegevens.nl/nl
Mr Aleid WOLFSEN, Chairman of Autoriteit Persoonsgegevens

Poland
The Bureau of the Inspector General for the Protection of Personal Data - GIODO ul. Stawki 2

00-193 Warsaw
Tel. +48 22 53 10 440
Fax +48 22 53 10 441
e-mail: kancelaria@giodo.gov.pl; desiwm@giodo.gov.pl
Website: http://www.giodo.gov.pl/
Ms Edyta BIELAK-JOMAA, Inspector General for the Protection of Personal Data

Portugal

Comissão Nacional de Protecção de Dados - CNPD
R. de São. Bento, 148-3°
1200-821 Lisboa
Tel. +351 21 392 84 00
Fax +351 21 397 68 32
e-mail: geral@cnpd.pt
Website: http://www.cnpd.pt/
Ms Filipa CALVÃO, President, Comissão Nacional de Protecção de Dados

Romania

The National Supervisory Authority for Personal Data Processing
President: Mrs Ancuţa Gianina Opre
B-dul Magheru 28–30
Sector 1, BUCUREŞTI
Tel. +40 21 252 5599
Fax +40 21 252 5757
e-mail: anspdcp@dataprotection.ro
Website: http://www.dataprotection.ro/
Ms Ancuţa Gianina OPRE, President of the National Supervisory Authority for Personal Data Processing
Ms Raluca POPA, Department of International Affairs

Slovakia

Office for Personal Data Protection of the Slovak Republic
Hraničná 12
820 07 Bratislava 27
Tel. + 421 2 32 31 32 14
Fax + 421 2 32 31 32 34
e-mail: statny.dozor@pdp.gov.sk
Website: http://www.dataprotection.gov.sk/

Ms Soňa PŐTHEOVÁ, President of the Office for Personal Data Protection of the Slovak Republic
Mr Jozef DUDÁŠ, Vice President

Slovenia

Information Commissioner
Ms Mojca Prelesnik
Zaloška 59
1000 Ljubljana
Tel. +386 1 230 9730
Fax +386 1 230 9778
e-mail: gp.ip@ip-rs.si
Website: https://www.ip-rs.si/
Ms Mojca PRELESNIK, Information Commissioner of the Republic of Slovenia

Spain

Agencia de Protección de Datos
C/Jorge Juan, 6
28001 Madrid
Tel. +34 91399 6200
Fax +34 91455 5699
e-mail: internacional@agpd.es
Website: https://www.agpd.es/
Ms María del Mar España Martí, Director of the Spanish Data Protection Agency
Mr Rafael GARCIA GOZALO

Sweden

Datainspektionen
Drottninggatan 29
5th Floor
Box 8114
104 20 Stockholm
Tel. +46 8 657 6100
Fax +46 8 652 8652
e-mail: datainspektionen@datainspektionen.se
Website: http://www.datainspektionen.se/

Ms Kristina SVAHN STARRSJÖ, Director General of the Data Inspection Board
Mr Hans-Olof LINDBLOM, Chief Legal Adviser

United Kingdom
The Information Commissioner's Office
Water Lane, Wycliffe House
Wilmslow - Cheshire SK9 5AF
Tel. +44 1625 545 745
e-mail: international.team@ico.org.uk
Website: https://ico.org.uk
Ms Elizabeth DENHAM, Information Commissioner
Mr Steve WOOD, Deputy Commissioner

European Data Protection Supervisor: Giovani Butarelli
Rue Wiertz 60
1047 Bruxelles/Brussel
Office: Rue Montoyer 63, 6th floor
Tel. +32 2 283 19 00
Fax +32 2 283 19 50
e-mail: edps@edps.europa.eu
Website: http://www.edps.europa.eu/EDPSWEB/

Chapter 3: Information Systems Management and the GDPR

At a Glance:

- ❖ Information Systems and their roles within the organisation
- ❖ Management Information System and major theories
- ❖ Data Flow Mapping: Use and Techniques
- ❖ The Data Controller and the Data Processor

Learning Objectives: *Students should be able to...*

- ❖ Explain the aspects of Management Information Systems: people, technology, and processes.
- ❖ Interrelate the importance of the GDPR to the 5 main components of Information Systems Management.
- ❖ Understand who is a Data Controller and a Data Processor, and the difference between the two.

Key Terms

1) Information Systems
2) Work Systems Theory
3) Cognitive Theory
4) Task-Technology Theory
5) Data Controller
6) Data Processor
7) Data Flow Mapping
8) Data Processing

Introduction

When an organisation deals in data of any kind, it requires a managerial framework that can organise the data into information. Information is what people speak, interpret, or analyse when they take data and organise it in a way that gives it meaning. With the high, and increasing, volumes of data that pass through the internet, there is a vacancy for organisations and institutions that can translate that data into a compressive set of information.

The importance of the GDPR is that it gives those organisations and institutions the responsibility of treating that sensitive data, with the respect it deserves. The information that is produced by the data sets are organised by the organisation into what is known as, quite evidently, an information system. The system is a set of complementary networks, that collect and analyse distributed sets of data. Business information systems not only rely on the processes of the technology involved, such as the hardware and the software, but also on the people within the organisation who interact with the technology in a way that facilitates business processes. Within the subject of Business Information Systems there is a set boundary of elements within a system, such as the outputs, users, storage, etc.

Types of Information Systems (IS) vary however, most are comprised of similar elements that will be outlined in detail in this chapter. These elements are the 5 Components; software, hardware, data, people, and procedures, and form the basic framework describing computer-based IS. The main types of IS that are most relevant to the GDPR are the Data Management Systems (DMS), more precisely; those that deal with personal data. Distinctions are sometimes made between business processes and computer networks, as separate systems, and although there is some difference between the two, it is important to remember that, the technological side interacts with the people who are part of the business process. The impact of this interaction, relative to the underlying principles of the GDPR, must be understood for the successful implementation of any compliance strategy.

Lastly, information systems provide an important and flexible resource for executives and leaders of organisations and institutions in the managerial decision-making process. This resource has created chief positions in many organisations directly dealing with the management of data, this role is known as the Chief Information Officer (CIO) and a Chief Information Security Officer (CISO), for the protection of information systems. The GDPR specifies a new role to incorporate data protection duties, the Data Protection Officer (DPO). The roles and further detailing of the DPO will be covered in chapter 7.

Information Systems in Organisations

Information Systems (IS) are a collection of complementary networks, such as technological networks, that interact with people to facilitate the business process. This section will analyse the elements of the information systems (mainly computer-based IS) in greater detail, how they relate with each other, security, and the GDPR. These five main components are the ingredients to the recipe that is information systems.

1) *Hardware:* These are the physical assets of the information systems. These are things such as the CPU (central processing unit) and the supporting devices (keyboards, monitors, storage equipment). The main pieces of hardware concerned with data protection are the storage devices; those that hold the sensitive data, such as the servers and their physical location.
2) *Software:* This is the supporting element of the hardware. Software tells the hardware what to do, having the nature of a symbiotic relationship; software could not function without something to interface with, in this case the hardware. The software contained in the hardware are the programs that run or the operating system of the hardware. The software can be viewed as the information system within the hardware that directs the data in the circuitry that produces useful/useable information.
3) *Data:* Information systems, including those that are not related to business, require inputs in a raw form, that can be organised; that is a set of diasporic truths that exist to be translated by the person that experiences them. For the information system this resource is called data; the bits of information that are manipulated by the software. Data is stored in both physical and non-physical locations, the spirit of the GDPR sets the importance on the proper management of sensitive personal data that is stored.

4) *Procedures:* Procedures are the processes, a set of steps, to be followed based on the governing policies of the organisation or institution. They act as a guideline to the operating of the computer system, this could mean both limitations placed on how employees use certain programs, or the code of conduct for people operating the facilities of the organisation. Procedures are akin to the software of people and processes, where the latter two would be the hardware.
5) *People:* This element of the information systems is the most fundamental in the system. Every information system requires people to firstly make it meaningful, and most importantly, useful. The people of an information system determine the success of the system. The people are primarily the staff that operate the hardware and software, the maintainers and organisers of the data, and the end-users or customers.

The data is the connecting force behind the information system, it is only when people interact with the other elements does the data become information.

Processes and Essential Systems

The two added features of an information system are the processes and essential systems that are a collection of important, but not directly related, elements that influences business activities. The essential system involves the environment that the core information system resides in, that is the inputs of the information system that results in an output or the purpose of the system.

An example of this could be a marketing firm dealing in mobile phones, it resides in the environment of the mobile market. Its inputs rely on consumer preferences, technological developments in mobile phones, social media habits of the customers, and the needs of the company's shareholders; these are a few rudimentary examples but will help define the environment. These inputs are then assembled and organised by the nuclear information system that resides within the environment to produce a certain output, or the purpose for the existence of the information system, in the case of the marketing firm to sell the consumer a phone and generating revenue for the firm.

Below is an illustration (Figure 3.1) of the extended information system, with the included essential environment, developed by Jack Zheng (2009). This is a basic view of the work systems theory in MIS, later discussed in greater detail.

Figure 3.1: Extended Information System

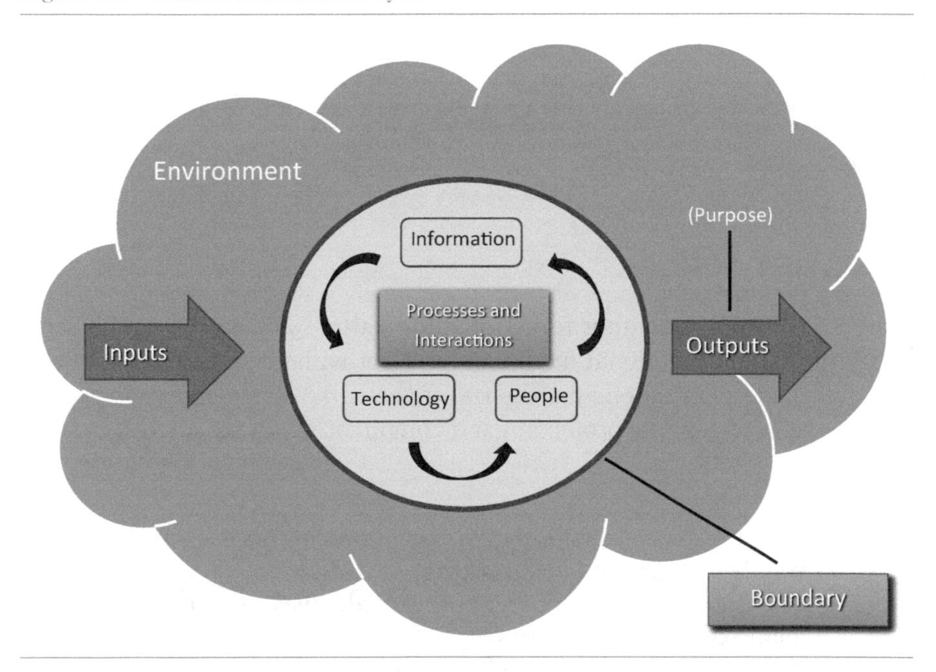

Types of Information Systems

Traditional models of information systems, constructed in the 1980's, were derived from the prevailing hierarchical structures within organisations at the time. The hierarchical system follows a pyramid structure, with executive information systems forming the cap of the pyramid, supported by the decision-making support systems, beneath which lies the management of information systems function, with the transactional processing systems forming the bottom of the structure (as shown in figure 3.2). The evolving nature of technology has left the traditional system of hierarchical organisation somewhat redundant, although it is still a useful tool in understanding the decision-making process within an organisation's information system.

Figure 3.2: Hierarchical system for information management

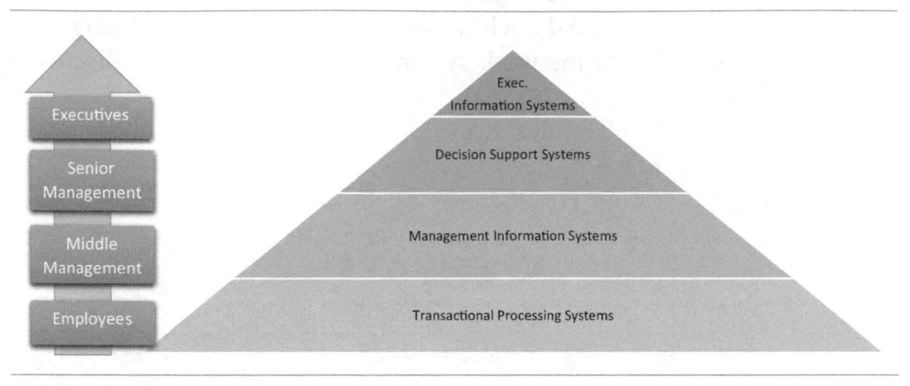

The newly emerging forms of information systems can no longer easily fit into the traditional model, at least not without some degree of agility within the organisations managerial framework. A few examples which tie into the need for data protection and compliance with the GDPR are:

- Web Development (web tracking, cookies, etc.)
- Data Storage (such as warehouses, hardcopies, server farms, or could-based virtual servers)
- Automatic Information and Decision Processing (Art 22)
- Search Engines
- Organisational Systems (Human Resources, IT, Marketing, etc.)
- Third Parties

The common denominator of the examples given above is that, in a modern sense, they are all computer-based information systems. These systems use technology to facilitate the tasks within the organisation. The components of a computer-based information system are mentioned in the section above titled information systems in organisations. The components of the information systems are based on a network which form the platform for the information system (hardware, software, data).

Information Management

What is IM

The management of information systems, or information management, is the key to compliance with the GDPR, knowing the types of information systems and how they operate and interrelate within an organisation is half the battle. The management of information concerns the flows of organisational activity; how the organisation acquires the information from the multitude of sources, and how the organisation stores and distributes that information to its end users and to those who require it. It also concerns the disposal, deletion, or archiving of that information and is most important when dealing with personal data.

Stakeholders

The management cycle of information within the organisation involves the stakeholders, employment that deals with quality control, accessibility, and use of the information, the security personnel that deal with the safe storage and disposal of information, and lastly the executive branch that utilizes it for managerial decision-making. In the context of the GDPR the stakeholders are the primary area of concern, as it is their personal data the legislation requires the organisation to protect. The stakeholders also have the right to change or delete the data that is stored, that regards them. This fact alone requires organisation dealing in personal data to have a strong framework for the management of information.

Information management does not differ much from traditional concepts of management. It still requires standard management concepts, such as the organisation, structuring, processing, controlling, evaluation, and reporting of the respected information to the correct department or leader. If these simple guidelines are met the system can fulfil its purpose (output) and allows the information to gain value by presenting it to the correct group of individuals. The GDPR only stresses the importance of the correct protections to be put in place, so that the sensitive data does not fall into the wrong hands or that it is not mismanaged in a way that causes damage, material or otherwise, to the data subject.

Ultimately the spirit of the legislation does not pose an obstacle to the success of an organisation where the availability of data is a key component to the prosperity of the organisation, but rather gifts the opportunity of a perspective change on the way data is collected and processed, in an increasingly complacent world of big data.

This section will outline a brief history of data management, the main functions of information systems management, and the main theories of information management. It is intended as a very brief introduction to the application and scope of computer-based information systems and the GDPR's relation to its process and operation.

Data Management through the Ages

Management of data, before the internet age, largely consisted of management through the use punch cards, magnetic tapes, and other forms of record-storing media. The life-cycle of these systems has not deviated much from what was the norm in the 1970's, both management frameworks required the acquisition, distribution, maintenance, and deletion/disposal of the data. In the late 1980's the world was transformed by personal computers and the potential for information technology to impact business processes, began to rise. E-mail revolutionised the speed and medium by which information was sent and the storing of data became much easier, as the size of computer chips decreased.

The development of information systems changed how we do business, as technology is now fully integrated into the business process as a means of value creation, rather than a function of the business activity. This development created a specialised category of jobs, where the management of information systems became highly strategic and required more senior positions in management. The ability to manage information systems is an indicator of business success and can be attributed to the capacity for the organisation to align itself with technological and business change.

A useful way to look at the change in business information systems is to examine the eras of modern management information systems. Although record keeping of information dates back to the use of ledgers (used in accounting for example) the modern computer-based information systems can be divided into 5 eras outlined in Laudon's (2010) work.

---

1) *First Era: Mainframes and Minicomputers*. During this period one computer would take the space of an entire room; basically, large calculators. In this period IBM dominated the space since they supplied the platform for the system. As the size of computer

components were reduced, businesses were able to integrate mini-computers into their activities, moving the business process into the second era.

2) *Second Era:* ***Personal Computers.*** In the 1960's the microprocessor began to compete with mainframe computers. The late 1970's saw microprocessors develop into personal computers; this move decentralized the power from data centres to offices and personal enterprise spaces. The low-cost of production for personal computers created the mass market commodities we see today.
3) *Third Era:* ***Client/Server Networks.*** It wasn't until the third era, that computer systems and information sharing became so complex, that they required most end-users, in this case employees and customers, to utilize the system, in order fulfil their function within the organisation. At this point the systems were connected through a network where all the information was shared through an intranet; similar to the internet, but rather a private network utilized by individuals within the organisation.
4) *Fourth Era:* ***Enterprise Computing.*** High-speed networks now connected all the intra-organisational software into enterprise application software (EAS). This software is a platform used by organisations, tailored to the business's needs, rather than individual use. This type of information system is used by organisations of all types.
5) *Fifth Era:* ***Cloud Computing.*** The current era is dominated by the rise of cloud computing. Cloud computing revolutionised the way business deals with information; it transformed the limitation of data storage and delivery of applications through the application of the latest developments in networking technology. Cloud computing opened access for management, allowing increased levels of information transfer through wi-fi connectedness and high speed mobile networks.

The transformative nature of information systems leaves a lot to be desired when it comes to the adequate protection of the data it handles; which in most industry sectors is poor. The GDPR creates clear requirements for data protection in the ever-evolving arena of information systems, but it is up to organisations and us, as citizens of the digital space, to change perspective on how we view data and our relationship with those organisations that process it.

Functions of Information Management

As discussed in the previous section the primary function of management information systems is to translate data into a useable information to facilitate the business process. There are a few steps involved in life cycle of an information system described below.

---

1) *Data Capturing:* this process involves the gathering of the data. How this is achieved largely depend on what the organisation does. For example, a social media site captures data through the user's utilisation of the platform, it is community driven and dependent on the user's compliance with the terms of service/consent. The data is gathered through various sources, such as customer touch points like stores or e-commerce websites or the output of other information systems like sales, market analysis, supply chain operation, etc.
2) *Data Storage:* the captured data is then stored on a database, be it physical or virtual. The storage system can be a hardcopy of any sensitive data held in folders within the physical location of the organisation, or more commonly, on a virtual server within the cloud. Data storage can be complicated when dealing with many employees (50+) as the GDPR requires a data flow map (discussed later in the chapter) for all sensitive data the organisation holds, including who has access to the data.
3) *Data Processing:* this function of MIS is at the core of what makes information systems useful. When the data is collected and then stored, the conversion of the raw data into meaningful information is what is known as data processing, without this process the raw data has no function. Data processing is generally done by an analyst, who then presents the data (through distribution) to the relevant authority at the right time, which may give the organisation a competitive advantage.
4) *Data and Information Distribution:* this aspect of MIS refers to the delivery of the information processed to the right person, at the right time. The information can be delivered in a variety of ways from email and files, to hardcopies such as reports (weekly, annual etc.) or video/auditory messages. The distribution of information is useful for management decision-making as it can facilitate comparisons with competitors or used in customer or production analysis.

5) *Predictions and Forecasting:* the sum of the functions above presents a useful tool for data management in business; the ability to forecast for potential opportunities, threats, and/or other events that may affect business activities. The business can make predictions using historical data as a basis for the research.

These are a few examples of the functions of management within an information system, but ultimately; the information system provides connectivity through technology and people, improves the managerial decision-making process, and improves business efficiency.

Information Systems Theory

Information systems theory is largely derived from traditional organisational theories and behavioural theories of management. The classical perspective was birthed during the period of the industrial revolution. It revolved around the efficiency of production and was output oriented, with little regard to the behavioural attributes of the employees.

The classical theories created a scientific approach to management where components of the production process are broken down into silos and scrutinized, to facilitate improvements within the production cycle. This later lead to the bureaucratic approach of management which created a hierarchical structure within the managerial framework by dividing tasks based on authority. The classical theories may be a little outdated, but the influence remains, creating an opportunity for organisations to implement frameworks that allow for the transition (or integration) of new information systems theories of management.

Newer theories of management incorporate classical theories (hierarchical etc.) with behavioural approaches. In the early 20th century, management research showed that, staff behaviours, attitudes, and motivations play a significant role in the success of the framework and the success of an organisation. It is important to realise that people are at the heart of the organisation and that the motivations and behaviour of those people effect the operations of the business.

Systems theory is the basis for the academic study of information systems management. The theory outlines the belief that all organisational components are interrelated. The interrelated components form what is known as the system and that alteration in one component may affect the function of all the others, positively or negatively. The system is in a continual state of flux around a dynamic equilibrium point, and it adapts to changes in the environment and the circumstances it is in; such as market activities, new regulation such as the GDPR, environmental changes, etc. There are more detailed theories within the systems theory that pertain to specific attributes of the systems approach, the two main ones are outlined below.

Cognitive theories

Cognitive theories are categorised under cognitive fit and cognitive dissonance theories. The theory holds that the way information is presented affects the performance of the task. First developed by Iris Vessey in 1991, the theory is based on the importance of information presentation; the format can influence task performance for the individual user. In the academic research conducted on cognitive fit there were key performance differences among individuals who were presented with a range of formats; graphs, schematic faces, and tables. The theory highlights the need to match representations to the tasks, facilitating the problem solving-process.

There has been a wide amount of research on the topic, with many findings showing that the form in which information is represented enhances the task performance when there is a cognitive fit. On the other hand, cognitive dissonance occurs when the attitudes and behaviours of the internal organs of an organisation, such as the employee or the various managerial branches, are not in alignment. This can create a rift between the actions and the intentions behind the operation of the information systems. There should be a concerted effort by management to try and eliminate any inconsistencies between the attitudes and behaviours of the organisation.

Task-technology

Task-technology fit theory, proposes that the capabilities of information technology (IT) must provide a best match to the user's task, so as to have a positive impact on the business process. Goodhue and Thompson's theory, formulated in 1995, measures the task-technology fit based on 8 factors. The measurement was based on a 10-question survey surrounding each factor: quality, locatable, authorisation, compatibility, ease of use/training, systems reliability, relationship with user, and production timelines.

The resultant measure of task-technology fit is used in conjunction with the utilization, to ascertain the impact of technology on individual performance within the information system. Unlike cognitive theories, task-technology fit examines the technological impact, rather than the form of presentation of information, as a measure of performance. The theory can both be used to analyse individual performance or on a group level with the analogous model of Zigurs and Buckland (1998).

Work systems theory

Many of the theories within the IS sphere rely on the contribution of the user to evaluate the performance of the system. The work systems theory looks at the system from a wholistic view, it does not view specific parts of the IS as having an impact on the performance of the system, rather, it views the system as a work system by default, with technology being a component of the system as a whole. Essentially the participants (the people) and the machinery work together using technology, information, and various other resources to produce the intended output, in a commercial sense; products and services. An example of a fully automated work system is a search engine, designed by people, the machine's components then operate on its own to produce the output.

The work systems life-cycle model maps the work system as a dynamic IS, which changes over time. The understanding, that the life-cycle of a work system can change both internally (planned change) or externally (unplanned or emergent change) makes the work system theory agile. The planned changes can come from projects within the IS that initiate a new development or implementation of new policy (like the GDPR). The unplanned changes arise when the environment the IS sits in requires the work system to adapt or workaround certain factors, such as competition or new regulation. In the case of the GDPR this change can be either planned or unplanned, depending on the organisation's approach to compliance.

Which theory works?

The theories mentioned above are a small percentage of many within the study of information systems. The important thing to realise is that there is no single correct theory, many of the theories do well when related with one another. Ultimately, not every organisation is going to adapt to one specific model and dependent on a variety of factors, one might work better than another. The appendix of this chapter contains a few diagrams related to each theory listed.

Data Flow Mapping

A data flow map is a diagram which shows the journey of data through the organisation. Mapping the flows of data between points within and outside the system improves decision-making. Data flow mapping is a key 1st step in the preparing of a Data Protection Impact Assessment (DPIA), the full details of a DPIA will be discussed in chapter 7. Data mapping is essential to understanding how the information system of the organisation is operating; it outlines the areas in which the organisation holds any sensitive data and the lifecycle of that data. To comply with the GDPR, the organisation should walkthrough the information lifecycle; finding what personal data is held, where the data goes, who utilises the data, and identify the intended use of the data up until the destruction of that data. Destruction is a key point within the information lifecycle as the regulation also stipulates that data should not be held any longer than its intended use.

The point of this within the context of the GDPR is to ensure that the natural persons (data subject) is consulted on the practical implication of the use of their data and any future planed use of the data. This is an important facet of data flow mapping as holding data just because it might be useful later, is prohibited under the GDPR, the organisation must be clear that any data it collects has potential future use before it is collected.

The components of a data flow map under the GDPR revolve, specifically, around personal data. The purpose of the legislation is to protect the data subject and their personal data; not the entity that processes it. When first preparing a data flow map it is necessary to identify what personal data the organisation holds or is collecting. Personal data is any data that can identify a natural person: names, addresses, ID numbers, and less obviously the digital identifiers such as IP addresses, health data, biometrics and much more. Chapter 1 describes in full detail what constitutes personal data. A preliminary guide to overcoming the challenges of data flow mapping begins with:

1) *Identify Personal Data:* What personal data is the organisation collecting, from where are they collecting that data, and what are the reasons for the collection of that data; the purpose of processing.
2) *Technical and Organisational Safeguards:* What safeguards are in place to protect the data from either accidental or malicious loss,

destruction, or to ensure the availability of the data. Is the data being handled properly?

3) *Legal and Regulatory Framework:* Understanding the legal and regulatory obligations around the handling of data, so that the processing of the data is compliant with the regulation.

These few steps are key to building trust with the data subjects and with the regulatory bodies of the GDPR; it is a simple checklist for understanding and overcoming the challenges of an effective data flow map.

Four elements of Data Mapping

The Four key elements of data flow mapping are the:

Data item: name, health data, criminal records, etc.
Format held in: hardcopy, digital, virtual servers, etc.
Location of the data: in the office, on a data base, the cloud, third parties, etc.
Method of transfer: through the post, email, over social media, etc.

This part is especially important as the way the data is moved changes the risk associated with the loss or destruction of that data. These key elements will assist in effectively mapping the organisation's data. In summary what data flow mapping is trying to accomplish is:

- Understanding the organisational information flow: the transfer of data from one location to another.
- Describing the information flow: the information life-cycle utilised as a tool to distinguish unforeseen or unintended uses of the data, ensuring that the data subjects are informed on the practical application of the use of their data, and to outline potential future uses of the data given the lawful basis for processing.
- Identify the key components of personal data: the data items held, the format it is held in, the transfer methods utilised, the storage location of the held data, the accountability of the data controller and or processor, who has access to the data, and has adequate protection been applied.

Data Flow Mapping Techniques

There are a few ways an organisation can achieve an effective data flow map. It relies on the strengths of the organisation's personnel and the management framework used, since a technique that might work for one organisation may not suit the needs of another. If the organisation has some certified managerial practice surrounding data protection, such as the various ISO certified standards (discussed in detail in chapter 9), it may already have a system in place for data flow mapping.

In reality many organisations seldom document their workflows, let alone their data flow maps. Fortunately, the GDPR only requires that the organisation map the processing of personal data. There are many ways an organisation can document the data flow of the personal data they hold. The focus of a data flow map is to view the reality of the flow of data and not what happens in theory. A chart looking at the various techniques of data flow mapping can be seen below in Figure 3.3.

Figure 3.3: Data Mapping Techniques

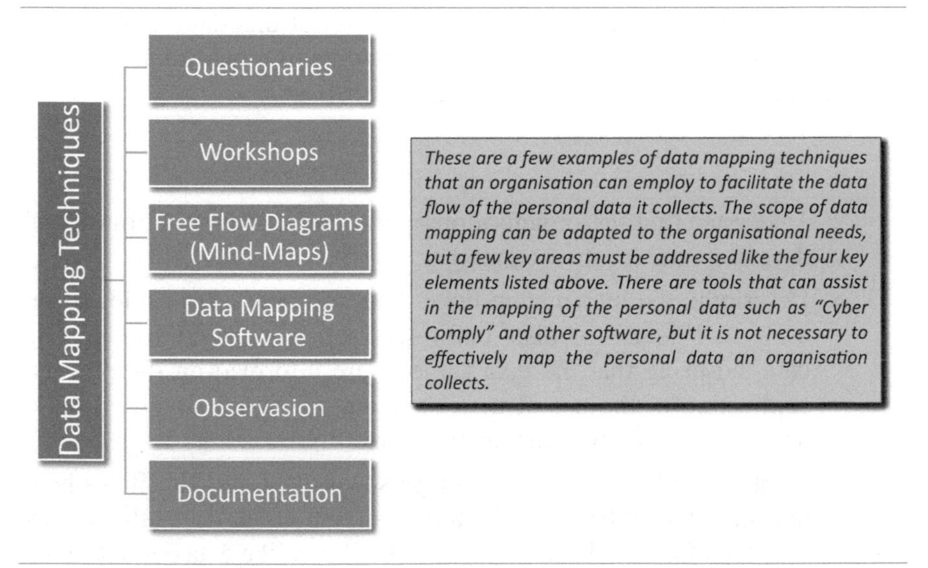

The purpose of performing a data flow mapping exercise is to identify where the data enters the system, the inputs, where it flows within the system, and where it exits the system, outputs. As described in the previous sections, an information system, in this case a work system that deals with data, is a culmination of inputs and outputs that fulfil the intended purpose of the system. Within a data flow map relevant to the GDPR however, it is important to scrutinize the inputs, such as:

- How is the personal data being collected and stored, e.g. via email, phone call, online, surveys etc.
- What is the personal data being used for?
- Accountability; who is responsible for the collection of the data are they using adequate safeguards
- Where is the personal data being stored/located; on a database, third party servers, cloud-based storage on virtual servers etc.
- Is the personal data being shared with third parties or suppliers and are they using effective data mapping techniques?
- Does the system share or communicate the information with other systems, for example automated processing systems?
- Who has access to the information, such as employees, do they have clearance to access that information?

The data flow map is a useful tool in assisting with data protection, but what does it look like in action? Below is an example of a data flow map for a marketing firm that collects personal data through online surveys, call centres, and from third party sources (Figure 3.4). In the GDPR the marketing firm would be considered a data processor, described in greater detail in the next section.

The key reason an organisation should create a well-developed data flow map is to facilitate compliance to the requirements of the GDPR. One example where this compliance proves useful is the DSAR; Data Subject Access Request (detailed in chapter 7), where a data subject exercises their right to access their personal information. An organisation cannot locate the personal data requested if it does not know where the data is located, and this inability to grant a DSAR is considered a breach of the GDPR. Furthermore, if the data subject exercises their right to erasure, the data flow map will again assist in the task of locating and then deleting the relevant data. Ultimately it is good practice to get the data flow map right, for the sake of the individual and for greater efficiency in meeting governance requirements for the organisation.

Figure 3.4: Simple data flow map

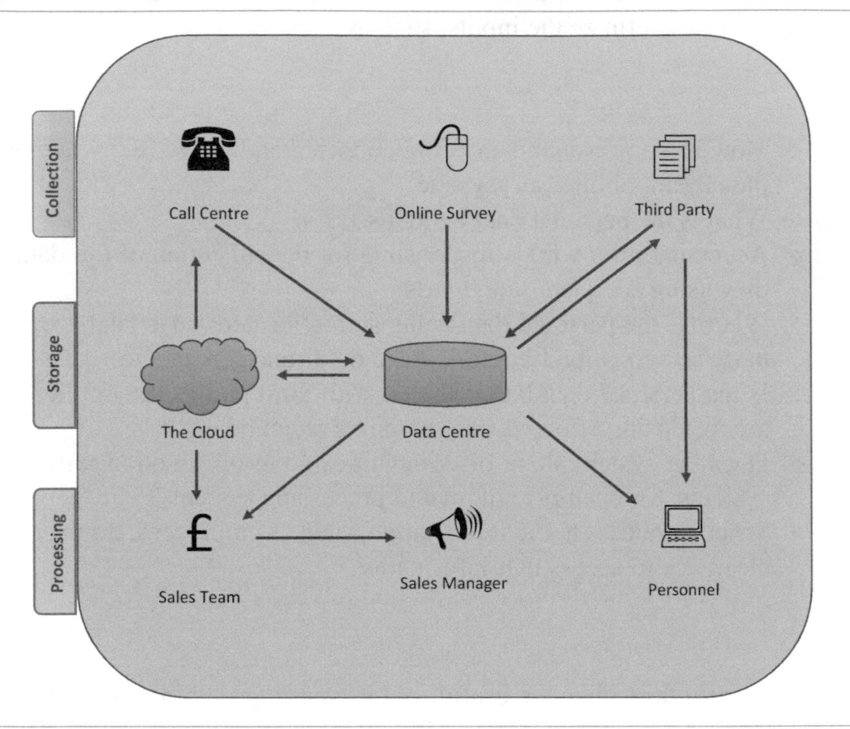

Data Controller and Data Processor

The GDPR describes the processing of data as, "obtaining, recording, or holding the information or data or carrying out any operation or set of operations on the information or data"[1], these can include the following:

- Adapting, altering, or organising the data or information.
- The retrieving, consulting, or use of personal data or information.
- The disclosing of information or personal data through; transmission, dissemination, or otherwise making available.
- Alignment, combination, erasure, destruction, or blocking of the information or personal data.

[1] The ICO's description of data processing under the DPA (data protection act 1998).

Data Controller

The Data Controller (DC), within the GDPR, is the person or entity that holds the data and shares it with the data processor; in effect the controller is responsible for the data which it distributed to processors. The role of the DC involves assigning the purpose behind the collection and processing of the personal data and the means of processing the personal data. The DC is tasked with the safekeeping of the data, and although the data processor also carries responsibility for the security of that data, it is the DC who carries the highest degree of responsibility and is the first to be held liable for a data breach. It is the responsibility of the DC to ensure that any third-party processors, appointed by them, are compliant with data protection regulations.

The primary tasks of the data controller are as follows:

- The data controller must ensure they comply with the regulation (fair processing of personal data.
- The data controller must obtain the data fairly (lawful basis and they must inform the data subject) and to utilise it only for its intended purpose.
- The data controller must implement adequate technical and organisational safeguards.
- The data controller must adequately manage any data processor that they employ.

Ultimately, the Data Controller exercises overall control, and use, for the processing of personal data and is held to be fully accountable for the actions of third-party processors of that data.

Data Processor

The Data Processor (DP) acts as an agent of the Data Controller, who is engaged in the processing of the data on behalf of the DC. The DP's obligations are to the Data Controller, although the DP still must comply with the GDPR. The tasks of the DP are based on the requirements of the DC;

however, these requirements must be fulfilled in accordance with the obligations of the GDPR. At a minimum the DP must adhere to the following:

- The Data Processor records the processing of the personal data and stores it with adequate technical and organisational safeguards.
- The Data Processor must inform the data controller and the relevant supervisory authority of any data breach that may have occurred.
- The data processor must obtain written permission by the Data controller for any third party or sub-contractor that uses the personal data.
- The Data Processor must appoint a Data Protection Officer if required (DPO, chapter 7)
- The Data Processor must comply with any data audit that is requested of them.

In summary the primary obligation of the Data Processor is to the Data Controller; who in turn allows the DP a certain degree of autonomy over the processing of the personal data within an agreed upon framework.

Distinguishing the Difference Between the Data Controller and the Data Processor

Before establishing the differences between the Data Controller and Processor, it is important to note that an organisation can act as a controller *and* a processor of data. The main legal differences between the two functions arise from the nature of obligations: The DC's obligations lie with the citizens of Europe, whilst the DP's obligations lie with the Data Controller for whom they do work. The reason a distinction is made between the controller and the processor is to recognise that not all organisations and institutions have the same degree of responsibility when it comes to the processing of personal data. In this case the DC must exercise the control and, by default, bear the responsibility for the processing of the personal data.

There are certain decisions that only the data controller can make, such as what personal data to collect, the lawful basis for doing so, which data items to collect (names, addresses, etc.), whether to disclose the data and to whom, how long should the personal data be held, etc. The data processor seldom will decide on these, as they are engaged by the data controller with the task of processing the data for its intended purpose.

For example, when an organisation outsources its recruitment scheme to a third-party HR company and requests that employee records be processed for the purpose of the organisation's recruitment program. The HR company's obligations are to the organisation employing them for their services, it is up to the Data Controller to make sure the third-party service provider complies, to not only its own requirement for the appropriate handling of the data but, to the regulations guidelines.

In the above example, *Buchanan Inc.* is the entity employing staff and thus becomes the data controller, since they are holding the potential employee's personal data. The third-party HR services, *Napier and Friends,* is then hired by *Buchanan Inc.* to process the employee records to facilitate the employment program of *Buchanan Inc.* The HR service provider, *Napier and Friends*, then becomes the data processor. Below is a chart describing the obligations and differences between the two companies; acting as a data controller and as a data processor (figure 3.5).

Figure 3.5: Data Controller v Data Processor

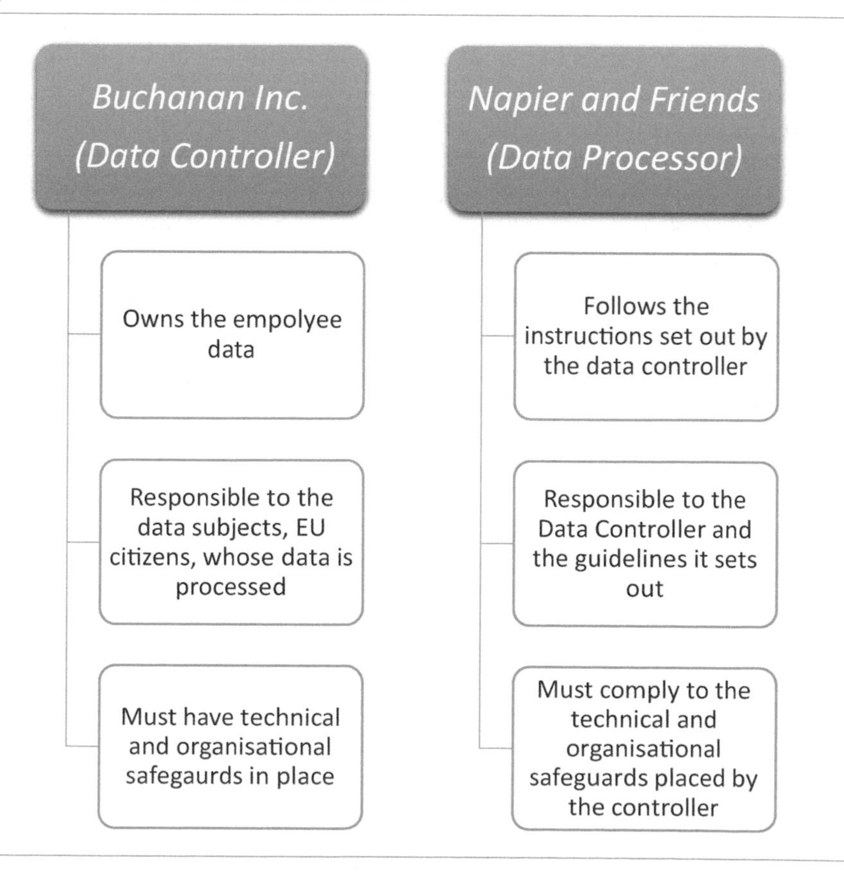

This chapter covered the basics of the internal and external operations of a business information system. The important thing to note about Information Systems, is that it should be examined as a flexible and adaptable set of processes that fit within an organisational framework. The better an organisation understands the environment in which it operates, the easier it is for the organisation or institution to comply with the regulation. The elements covered should give an introductory framework to effectively define the boundaries of an organisation's information system, but also the risks associated with failing to understand the management and processing of the system.

The ideal way to view an information system, in an organisation, is to understand how the technology, the hardware and software (in a computer-based IS), interact with the people to facilitate the business process. The many theories in the literature on information systems, should be viewed as a cognitive exercise to, further exemplify the practice of managing the flow of information and its interaction in a real-world scenario. The theories should also build upon the functions of the information system and the use case for everyday business activities and management. These basic concepts in management information systems should then aid in building an effective data flow map, with the understanding of the roles and tasks associated with the liable parties, the data controller and the data processor.

The GDPR is a tool that can assist in illuminating the significance and impact a well-managed information system can have on organisational efficiency. The spirit of the law gives not only the organisation, but the citizens of Europe, the needed perspective change on the way society and the individual see themselves in the digital space; that their rights are not to be subjugated purely due to the changed environment of daily life.

In the next chapter the theme will shift from the information systems within the organisation and its relevance to the GDPR, to the technical protection of data using cyber security tools.

Chapter Review

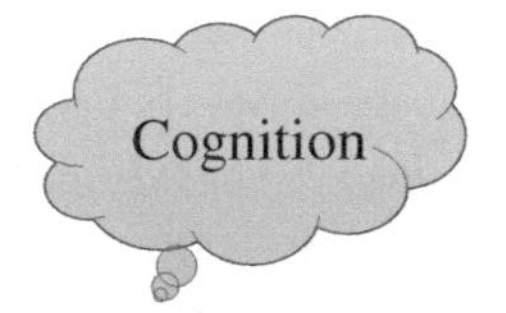

Summary of Key Points

1) A general view of information systems and their function.
2) Information systems within an organisation.
3) Information systems theory.
4) Data flow mapping techniques and application.
5) The roles and tasks of a data controller and data processor.

- ❖ Think about the functions of information systems and its management.
- ❖ Consider the liable parties, and the difference between a data controller and processor.
- ❖ How would you conduct and data flow map and a data audit?

References

Alter, S. (2006). Work systems and IT artifacts–does the definition matter? *Association for Information Systems (Volume 17, 2006), 17*, 299–313.

Alter, S. (2008). Defining information systems as work systems: Implications for the IS field. *European Journal of Information Systems, 17*(5), 448–469.

Alter, S. (2013). Work System Theory : Overview of Core Concepts, Extensions, and Challenges for the Future. *Journal of the Association for Information Systems, 14*(2), 72-121.

Bostrom, R., & Heinen, J. (1977). MIS Problems and Failures: A Socio – Technical Perspective, Part II: The Application of Socio – Technical Theory. *MIS Quarterly, 1*(4), 11.

Calder, A. (2017, October 5). *Data flow Mapping Under the EU GDPR*. Retrieved from IT Governance: https://www.itgovernance.co.uk/gdpr – data – mapping

Davis, F. (1985). *A Technology Acceptance Model for Empirically Testing New End-User Information Systems.*

Dishaw, M., & Strong, D. (1999). Extending the technology acceptance model with task-technology fit constructs. *Information and Management, 36*(1), 9–21.

Ferrara, P., & Spoto, F. (2018). Static analysis for GDPR compliance. *CEUR Workshop Proceedings, 2058.*

Goodhue, D., & Thompson, R. (1995). Task-Technology Fit and Individual Performance. *MIS Quarterly, 19*(2), 213.

Information Commissioners Office. (2017). *Key Definitions*. Retrieved from ICO.: https://ico.org.uk/for-organisations/guide – to – the – general – data – protection – regulation – gdpr/key – definitions/

Information Commissioners Office, ICO. (n.d.). *Data Controller and Data Processor: what the difference is and what the governance implications are*. Wilmslow: ICO.

Information, M. (2005). The Gordon B. Davis Symposium The Future of the Information Systems Academic Discipline:. *Information Systems Research, 28*(4), 1–8.

ITGovernance. (2017). *GDPR Penalties*. Retrieved from ITGovernance: https://www.itgovernance.co.uk/dpa – penalties

Tankard, C. (2016). What the GDPR means for businesses. *Network Security, 2016*(6), 5–8.

The European Parliment and The Council of the European Union. (2016). GDPR. *Official Journal of the European Union*.

Vessey, I. (1991). Cognitive Fit: A Theory Based Analysis of the Graphs Versus Tables Literature. *Decision Sciences, 22*(2), 219–240.

Work, B. (2002). Patterns of software quality management in TickIT certified firms. *European Journal of Information Systems, 11*(1), 61–73.

Chapter 4: Cyber Security and the GDPR

At a Glance:

- ❖ Introduction: Cyber security as a function of compliance
- ❖ The 3 P's: Privacy, Protection, and Processes
- ❖ Cyber-attacks: Malware *vs* Virus and Phishing
- ❖ Social Engineering
- ❖ Countermeasures: Encryption and its function

Learning Objectives: *Students should be able to...*

- ❖ Distinguish the importance of cyber-security as helpful tool in GDPR compliance.
- ❖ Define and give examples of types of cyber-attacks and to explain their purpose and damage potential.
- ❖ Identify protective measures against cyber-attacks and illustrate their application.
- ❖ Recognize the current issues for cyber security going forward.

Key Terms

1) Privacy
2) Protection
3) Processes
4) Malware
5) Virus
6) Phishing
7) Social Engineering
8) Encryption
9) Hashing
10) Asymmetric Encryption
11) Symmetric Encryption

Introduction

Cyber security is the disciplined protection of the confidentiality, integrity, and availability (CIA) of information. This includes the methods with which an organisation, or individual, protects computer systems from unwanted damage to hardware, software, and information. The scope of cyber security includes the protection against damage, to systems, and also protection from the misuse of network systems; resulting in misdirection or disruption of the services an organisation provides.

Cyber security is growing in importance and relevance, with society's increasing dependence on computer systems, such as the Internet, smart devices (IoT), and wireless networks, providing an ever-expanding range of attack vectors for cyber criminals. Unfortunately, the increasing dependence and use of computer systems has not correlated with increasing interest or use of cyber security methods. Data breaches have increased exponentially over the years between 2000 and 2018, with only 4% of those attacks achieved through bypassing encryption security.

There are some misconceptions about cyber security, a common one being that protection is provided solely through the use of software programming, on non-tangible assets. In reality systems operate on physical assets, where physical access is required for the operational longevity of those information systems. An example of this is a data centre; the location of which may need to be kept secret. If it processes sensitive information, any damage or infiltration could result in loss of availability to the data or unauthorized access and use of the data.

As the old proverb goes; *the road to hell is paved with good intention.* It is unlikely that the first pioneers of the internet, such as Tim Berners-Lee, Claude Shannon, Donald Davies, or Paul Baran, had the intention to build security into connected network systems. It was in the 1970's when the original cyber punks, such as Ron Rivest, Adi Shamir, and Leonard Adleman, founders of the RSA encryption method, developed new forms of cryptography for the encryption of information. The rise of encryption in modern times gave precedence to the importance of security for communication systems.

With the current state of the world's communication infrastructure there is a long, but not impossible, task of re-envisioning what it means to be a global citizen in an unsecure digital world, with the self-sovereign individual as the sole provider and owner of their own data.

Cyber Security as a Function of Compliance

Cyber security and the GDPR go hand in hand. A few of the GDPR's articles indicate the requirement for the adequate protection of personal data, through encryption and the use of pseudonymisation. Cyber security is more than just a tool for GDPR compliance, it goes beyond showing good faith with the regulators. It should be incorporated in the business culture for start-ups, established institutions, and organisations. The GDPR is fundamentally about the rights to, and the protections of, an individual's data; this is paramount in the spirit of the regulation.

Natural persons have the right to privacy, protection, and processing of personal data; the 3 P's of data security, and the building blocks of the digital citizen's rights. With the recent rise of whistle-blowers, such as Edward Snowden[1], there has been ongoing public debate surrounding the issue of privacy, especially in the context of national defence.

Privacy

Broadly defined, privacy is the ability to segregate or isolate oneself, or information relating to oneself; providing freedom for selective expression. The right to privacy involves two distinct elements; non-intervention, and protection. Surveillance, secret or otherwise, is an intervention which impacts an individual's privacy; therefore, where this surveillance occurs it should be with the subject's consent.

There is heated debate, on the subject of privacy in the digital space. The main argument is based around the potentially adverse effects to national security which arise from the rights to data protection. The question of how much privacy should be sacrificed, in order to protect the nation from potential cyber threats, or physical threats such as terrorist attacks, is biased. The correct question should be; what can be done to ensure the citizen's privacy *and* provide the access to information required for national security.

Cyber security personnel, governments and other relevant organisations that control and process data, should not know everything or have access to all data, about an individual to protect them adequately from cyber or physical threats. The same rights of the natural persons in the home should be extended to their digital life, as the hallmark of a developed cyber world.

[1]Revealed information, in 2013, on the mass surveillance of the public by government agencies.

With the continuing threat to the individual's digital privacy, there is a need for privacy ethics to be built into new and existing network structures. Within the GDPR privacy is a right for the natural person, and any sensitive data should be pseudonymised, as a measure of compliance. Essentially privacy entails the use of data masking techniques, so that only the data controller, the data processor, and the data subject, know the details of the data and so that anyone else viewing the data should not be able to identify the data subject, without the use of an encryption key or other important information required to access it.

Ultimately, privacy is an intrinsic human right, enshrined in law, that should not be ignored in cyber security. It is often stated that if you have nothing to hide then lack of privacy should be a non-issue, this concept is misguided. For example; the contact details of all the authors are in this book, yet if the login details for the social media accounts, of the readers of this book were requested by the authors, it would be quickly seen that people take their security details and the right to privacy to their content very seriously.

Protection

The second P, protection, lies with the institutions and organisations controlling and processing personal data. Adequate protection of sensitive data is a requirement for GDPR compliance. The data subject has the right to expect appropriate protective measures are taken to secure their personal data, although it is not the data subject's responsibility to make sure it occurs. The institution or organisation should protect sensitive data through the use of encryption and other relevant tools, in order to comply with the GDPR.

The primary reason for the protection of personal data is to prevent, and minimize, the effects of a data breach. Data breaches can occur in any scenario, from the malicious actions of attackers, to the accidental loss or destruction of personal data. The sensitivity of the data should inform the level of protection that is used, for example medical data, compared to data on an individual's interest in a certain type of music. In the context of the GDPR, highly sensitive data deserves greater protection, due to the scope of identifiable attributes held within them. Medical records contain details such as biological attributes, home addresses, and the types of health conditions the individual might have.

Reusing the example of musical taste, this data might not fully identify the natural persons, but marketing organisations in the industry would find this information very useful for marketing purposes. What the GDPR nominates is that even the misuse of what might be considered somewhat useless data, in the world of big data, is grounds for penalties. Big data is the astronomical mountain of information that is uploaded and downloaded from the internet, the IoT, and the back end of systems. Big data analytics is the use of computational systems to analyse trends, patterns, and associations of human behaviour that businesses often use to approach new opportunities and strategies.

So, what is important about big data? At first glance, information about an individual's taste in music, or general likes and dislikes, may not seem worth protecting but with enough data, anyone can be identified. Cryptographer Bruce Schneier, in his book *Data and Goliath,* stresses the need for greater protection in the cyber space for the individual. The way in which big data is processed can easily identify a person, purely based on interests, dislikes, location, and browsing patterns; all without the need of a sensitive data breach.

A simple example of this would be an advertising firm that collects meta-data to analyse consumer spending patterns. In the stereotypical sense, finding and identifying a specific person that is aged 18-25, in a university environment, that uses social media, and consumes streaming services, videogames, *and* who is also male, would be rather challenging as there would be hundreds if not thousands of men who fit that profile.

On the other hand, identifying someone through meta data, who drives a yellow convertible, is 93 years old, and who likes to play golf in Scotland, would be much simpler than the example with the university student, stated above. This example is used to illustrate that even without the names, addresses, and other personal details that may identify a specific individual, an analytics firm could still hold some key information about the person that can be used to identify them through the analysis of the meta-data.

As stated previously, what might not be worth protecting in the context of the GDPR is of greater importance in the context of big data analytics and its implication on digital civil liberties addressed within the GDPR. Although the use of big data analytics blurs the lines when addressing personally identifiable data, it is good business practice to minimize exposure by protecting data, minimizing the volume of data stored, and deleting any unnecessary data; avoiding what is known as a data lake. Above all, it is a measure of good faith; when compliance is met with an optimistic attitude towards data protection and its impact on society, from the perspective of regulatory institutions, such as the ICO.

Process

The final P is process: this entails the processing and use case for the data. If data were to be viewed as a commodity (it is often referred to as the new oil) then it can easily be seen as going through a production life cycle; from raw material to finished product. Raw data, in essence, is the bits of information that are constantly being uploaded and downloaded from network connected sources. The 1's and 0's that make up the structure of information is harvested, often by data analytics companies (which in the oil analogy exemplifies the refineries and drilling platforms), and then sold on as a form of useable information packets, for a multitude of purposes. Some of the more widely known purposes are for use in advertising, studying macro and micro-economic models of consumer and producer behaviour, or in a more sinister way, for the quelling political dissent or controlling the outcome of elections.

Process within the context of the GDPR, pertains to the topics of data minimisation and right to erasure. These two processes are quite different, in their implications for an organisation and their implementation, but have overlapping elements that assist in the right to erasure and reasoning for data minimisation. As discussed in chapter 1, data minimisation is the process in which the data an organisation holds are stripped, or quantifiably minimized, to only contain what is necessary to maintain efficient levels of service and availability. The process of data minimisation can assist an organisation in compliance when a data subject exercises their right to erasure.

The primary reason process is included in the three P's of data security, is the necessity for data controllers and data processors to know their data flows, data inventories, and information architecture. Knowing these systems well, makes it easier to detect vulnerabilities and potential attack vectors that can be exploited by malicious actors.

Finally, and arguably the most important, is the process of people within the organisation. Many cyber security experts will note that data attrition and leaks are caused within the organisation by employees who are not adequately trained in data management. Accidental loss is still considered, under the GDPR, as grounds for a data breach, dependent on variables such as level of data sensitivity, level of encryption used on the lost data, and availability issues surrounding the accidental destruction of the data.

An example would be, an employee loses a USB containing sensitive employment records. There are some factors that must be considered when such an event occurs, such as which organisational policy is involved and if the USB had the proper encryption protocol in place. These are some of the types of questions that should be asked when reporting to the supervisory authority during an incident (detailed in chapter 9).

Process in data security is a balance between applying technical safeguards on any hardware and software network systems that have been investigated to be vulnerable to attacks and making sure that employees within the organisation are trained in data management and privacy, as to minimise the probability of a breach from the accidental loss or destruction of sensitive data.

Cyber Attacks

Cyber-attacks are an offensive action taken against computer network systems and network infrastructures by an organisation, nation-states, groups, or societies. Cyber-attacks are the 3rd biggest security risk, globally, after the risks arising from extreme weather. The means by which cyber-attacks are executed are through the use of malicious acts, usually coming from unidentified sources, with the intent of stealing, altering, or destroying a target. Cyber-attacks can cause a great deal of damage to an organisation or institution. The costs of the damages will not be fully apparent, until the aftermath of an attack. Some of the costs to be considered are; network infrastructure damages, data loss and theft (the sensitivity of the data drastically changes the cost), reputational cost to the organisation, and market value, to name a few. A cyber security framework conducted by NIST[2], show that average cost of a cyber-attack can range between £2-3 million.

Cyber-attacks can be generally dissected into two categories: ones that target systems in order to disable or render a system completely unusable, and attacks with the intent of gaining unauthorized access to stored data. Attackers use various means to gain unauthorized access, the following section will discuss a few of those techniques.

[2]US Department of Commerce, National Institute of Standards and Technology: figure translated from dollar to pounds (March 2018).

Malware

Malware, also known as malicious software, is a blanket term for any software that is intrusive or hostile by nature. The types of malware most commonly used today are, Ransomware, Spyware, Computer Viruses, Trojans, Worms, and the most commonly seen for website hacking, SQL injections. Not all the types of malware will be covered in this text, but some examples will be shown in the case studies littered through the chapter. The behaviour of malware changes with the type of malware used, the intent of the attacker, and the delivery method. Much like cyber-attacks in general, malware can cause theft, destruction, unavailability of data, or render systems useless.

Although malware and other cyber-attacks have the potential to create devastation to network systems, in the majority of cases, malware is designed to target *any* system it comes into contact with. These attacks are generic and generally not targeted; the attackers cast a net, in the form of the malware, and see what comes back. The attackers design poorly written malware that intends to harvest any information it can access; most malware attacks can be stopped with the use of up to date anti-virus software.

Sony BMG's Rootkit Case Study

In 2005 an incident involving the use of rootkit software by manufacturers, came to light. The implementation of a deceptive copy protection rootkit, by Sony BMG, was discovered on over 22 million of the company's CDs.

The Sony CD would install a hidden file on the user's computer drive adding a form of Digital Rights Management (DRM) that would interfere with the computer operating system's CD copying capabilities.

The malicious programs were difficult to uninstall, with some cases showing that even when a user refused to accept the EULA (end-user licencing agreement) the program would still dial back to base.

This resulted in reports of user's personal data, in the form of private listening habits, being sent to Sony BMG's data centres.

Although Sony BMG denied its harm, the rootkit exposed many user's systems to vulnerabilities from cyber-attacks.

Sony BMG eventually recalled 10% of their products and addressed further issues with consumer settlements following a public outcry and a government investigation in 2005–2006.

Virus vs. Malware

There are a few common misconceptions about computer viruses and malware, they are often used interchangeably; this is incorrect as there are differences between the two.

A computer virus is designed to spread from the point of infection by replicating itself. It works just like a biological virus; it inserts a line of code into the system (cell) with the intent of replicating and spreading to other machines. The copied virus can be slightly different from the original making it difficult to protect against at times, much like the flu virus when replicated can be hereditarily different from the original, making vaccination against it difficult.

ILOVEYOU Computer Virus

In 2000, an infamous virus originating from the Philippines spread through email. The computer virus was held in an attachment file called "LOVE-LETTER-FOR-YOU.txt.vbs". The file extension "vbs" was hidden by default on most windows systems at the time so to the user it appeared as a regular text file. Opening the file would activate the Visual Basic script (a programming language developed by *Microsoft*, it was declared legacy in 2008). The worm, as these types of viruses are known, would then delete files and hide MP3 files from the user. The ILOVEYOU virus prompted legislative change in the Philippines as prior to the writing of the virus there were no laws surrounding the creating and spreading of viruses. The creators of the virus Reonel Ramones and Onel De Guzman had the charges against them, dropped by the state prosecutor in 2000. The virus was awarded the record for most virulent computer virus at the time, 2002.

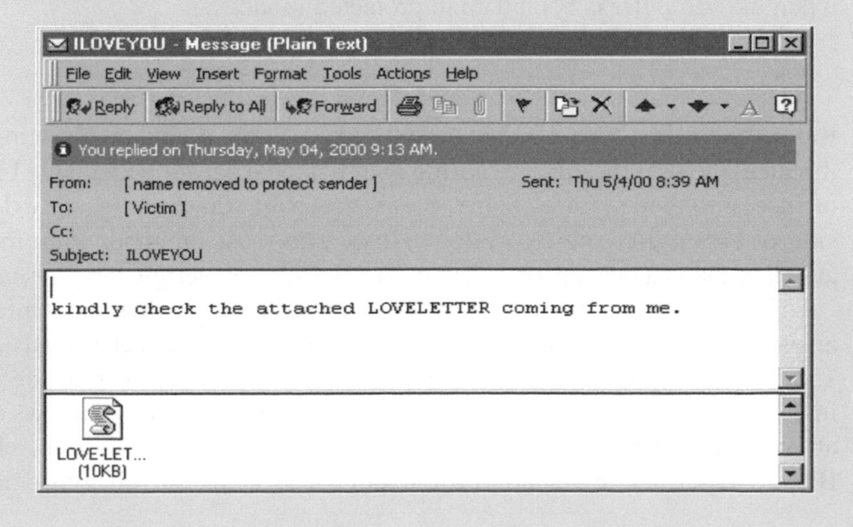

A computer virus which was originally implemented in a computer network, is a single line of code that has no host program unlike malware. The line of code can fuse with other programs or files and begin to replicate when specific conditions are met, such as a predetermined date and time, or when a certain program is opened by the victim. Malware, unlike viruses, are often a fully executable program that come in an *exe.* file. Malware does not rely on fusing itself with a file or a program, instead it can be passed around and is stand-alone. Malware can often carry a virus as part of its payload.

Rootkits and Bootkits

Rootkits, as in the example of the Sony BMG's case study, are a type of malware that affect the user modes of a computer. There are different levels of user access in a standard computer, from the lowest level guest access, to the highest-level administrator privilege. A *Ring 0* rootkit, or kernel-mode rootkit, is a one that uses the highest level of operating privilege. The kernel-mode rootkit replaces, or adds, code to the core of the operating system. This level of rootkit is difficult to write, but has complete unrestricted security access, and is therefore extremely valuable to an attacker. Kernel-mode rootkits are also difficult to detect, primarily due to their ability to operate at the same security level as the operating system itself.

Bootkits are more advanced versions of a rootkit, as they are able to infect the Master Boot Record (MBR) through infecting the start-up code. This infection allows for the bootkit to remain active even after a system reboot. It also has the ability to remain active in the system memory, even when the computer is booted up in protected mode.

Ransomware

Ransomware is a type of malware that locks down a network system, and threatens the user by either, exposing their personal data or perpetually blocking access to their systems; unless a ransom is paid. Ransomware is the darker side of encryption and its malicious use. The most infamous ransomware attack of the last decade was *WannaCry (aka Wannadecrypt, Cryptor etc).*

The *WannaCry* ransomware attack was a global cyber-attack that infected an estimated 200,000 computer systems worldwide. The attack spread across 150 countries, affecting the systems of public authorities, such as the NHS in the UK. The ransomware attacks targeted systems running Windows operating system. The malware encrypted *all* the host's files and demanded a Bitcoin payment (USD $300 worth) to unlock each system: threating deletion of files if demands weren't met in a timely manner. The attack started on the 12th May 2017 and lasted 4 days. Microsoft quickly patched their system after discovering a kill switch within the programming. The estimated cost from the attack ranged from, hundreds of millions to billions, of pounds. An example of the malware can be seen in figure 4.1.

Figure 4.1: *Wanacry* ransomware attack

This is what the victim of the WannaCry attack would be presented with when attempting to access their systems. The attackers were kind enough to inform the victim in the joys of cryptocurrencies, albeit a rather unpleasant crash course on how to buy and how to send Bitcoins.

Social Engineering

Social engineering in the context of information security is a form of psychological manipulation. It incorporates social means in the targeting and execution of an attack on a system, through non-technical vectors. Social engineering is a highly successful tool for cyber attackers as it relies on people's tendencies to trust each other. This generally involves a degree of social skills that allows attackers to harvest the necessary information required to bypass an organisation's security protocols. Even the most advanced security systems can be bypassed by simple social techniques.

Social engineering is usually a first step in an attacker's regime; it can at times function as a form of reconnaissance. Attackers will often scout out an organisation's vulnerabilities before carrying out the attack, through the use of social engineering. These techniques are the core tool for cyber-attackers, cyber security experts know that no matter how advanced a network's software or hardware security is, the real issue is trust, and the weakest link in the security chain is where trust is given freely, in this case people's willingness to trust one another.

It's a tough life to live when you can never trust anyone, and nor should you live like that (it's a loveless existence), but trust is something that should be earned, especially when entering any relationship. The same principles should be applied to organisational relationships, so it is important to have security practices built into the organisational culture, to avoid problems in the future. Most cyber incidences can be avoided with a balance between adequate personnel training and updated security systems.

The Psychology of Persuasion

Social engineering techniques are loosely tied to Robert Cialdini's work on persuasion. In his 1984 book *Influence: The Psychology of Persuasion,* Cialdini derived 6 key principles of persuasion. His work focussed mainly on marketing, specifically telemarketing and car dealerships, but has been adapted by cyber security firms and hackers looking to use social engineering techniques to exploit security vulnerabilities.

The six key principles analyse attributes of human behaviour and the decision-making process and has been primarily used as a marketing tool, however it is useful in understanding the role of social engineering in cyber security. The six principles are listed below:

1. *Reciprocity:* This principle describes people's likeliness to return favours. Marketers will often give out free samples in the hope that the customer receiving the sample will then "return the favour" by buying the product at a later date. In the context of social engineering this principle is exhibited when, for example, an attacker pretends to

be an employee that has forgotten their ID card and asks the receptionist, or other staff members, for assistance. This example illustrates the "today you, tomorrow me" scenario, where the receptionist or relevant staff member empathises with the attacker, and assumes that this could happen to anyone, including themselves, and is likely to help someone in need.

2. *Commitment/consistency:* This principle of influence plays on the individual's desire to up keep their self-image. People are more likely to pull through on an objective or idea if they commit, either in writing or verbally, especially if it aligns with their self-image. The commitment technique is used when attackers, through phishing techniques, attempt to gain the help of employees within the organisation, to access the system from the inside. For example, an attacker messages an employee in a large organisation and convinces them that by stealing company records they are doing the right thing. The attacker plays on the cultural perception that large organisations are out to take advantage of the average person. Playing on the victim's sense of morality engenders the idea that they are, in fact, doing a social good.
3. *Social Proof:* This is a principle of conformity. The technique demonstrates people's desire to copy what others are doing; to conform to the standard social practice, and not to deviate from what is social acceptable. One of the most influential experiments into conformity was research by social psychologist Solomon Asch. His experiments, at Colombia University, showed that most subjects would conform to the group decision, even when it went against what they knew to be true. The potential for attackers to exploit this technique is high; the victim falls to the notion that "everyone is doing this, so it must be alright".
4. *Authority:* As with social proof, the authority technique plays on people's willingness to listen to figures in authority. In the 1960's, social psychologist, Stanley Milgram conducted a series of controversial psychological experiments. The experiments, known as the Milgram Experiments, attempted to measure the willingness of participants to defer judgement and decision-making to a senior authority. The experiment required participants to administer increasing levels of electric shocks to an unseen "patient"; as directed by an authority figure, in this case, the researcher. The results where rather shocking: all participants administered electric shocks, of at least 300-volts, to

the screaming "patient", and 65% of participants administered a fatal shock of 450-volts, despite clear written warnings against doing so.

The power of authority to overcome normal decision-making is widely used in Japan when attackers impersonate an organisation's CEO. The corporate culture in Japan places great importance on the role of the CEO, and they are highly respected. An attacker can successfully gain unauthorized access by impersonating a senior manager and directing staff to carry out malicious activities. In other societies this attack may be executed by an individual masquerading as an IT technician or similar authority figure.

5. *Liking:* This principle is rather straight-forward; being a likeable person will get you further in life. Likeableness is a key factor in the ability to persuade others and is an important trait of successful sales people; and social engineering attackers. One method employed in a targeted attack, is the creation of a spoof social media account which aims to gain the trust and friendship of an employee. This is also known as pretexting, where attackers conduct research and reconnaissance to find a weak target to exploit.
6. *Scarcity:* Simply put, the scarcity technique is a persuasion tool that pressures individuals into purchasing or investing in an idea or product, due to high demand, which is often artificially created. The perceived level of scarcity generates interest in the idea, driven by the fear of "missing out". In product marketing this is seen in the "limited time offer" scenario that encourages the target into hasty decision-making, under time pressure.

Cialdini's theories, although helpful in understanding social engineering as a framework, are not completely transferable in the context of information security. As shown above scarcity is not transferable or applicable, in most cases, to social engineering. The principles should be viewed as a building block in understanding potential attack vectors and vulnerabilities in the organisational security structure and should serve as a lesson in the importance of information security as both a function of technical safeguards and the human psychological condition.

Dimensions of Social Engineering

Knowing all the social engineering techniques that are possible, would make it easy to implement countermeasures; in reality it is difficult. There are several ways an attacker can use social engineering techniques to achieve their goals. This section explores the different types of social engineering using the dimensional framework of social engineering of Terti and Vuorinen.

Terti and Vuorinen look to find ways to categorise the multitude of attack vectors capable in social engineering in a broad yet comprehensive manner. The framework they developed indicated three "dimensions" of social engineering, broken down into *Persuasion, Fabrication, and Data Gathering.* The three dimensions lay out a foundation for the types of social engineering techniques that an organisation or individual may face.

- *Persuasion:* This dimension of social engineering attempts to get the victim of the attack to comply with the attacker's desires, which are often inappropriate in the face of the organisation's security policy. Persuasion requires a degree of psychological manipulation of human emotion, and this could include exploiting people's trust, fears, greed, etc. Many elements of Cialdini's psychology of influence are exhibited in this dimension, as attackers will employe one or more principles to achieve their intended goal. Persuasion is the most direct approach of the three, as it requires the attacker's direct involvement with the victim to get them to comply with a request they would not otherwise do. The attacker actively engages with the target, utilizing the persuasion techniques established in Cialdini's theory.
- *Fabrication:* Techniques in fabrication involve fake identification, name-dropping, and jargon (see appendix for a full table of social engineering techniques). This dimension relies on the attacker presenting misleading cues to fool the target's interpretation of the situation. It uses well known theories, in psychology and sociology, showing that individuals interpretation of scenarios, is based on a multitude of factors, that are not obvious to the them. The attackers will "frame" the situation in a manner that presents itself to be normal to the victim.

  Unconscious factors, which are inherent in the victim, such as cultural norms, their own experience, and cues which are real or fake, affects their interpretation of the situation. Fabrication differs from persuasion in that it is more deceptive in nature (at least from the perspective of the target). The attacker takes a back seat and uses tools that "frame" the situation in a certain way so that the victim leads him or herself to achieve the goal set out by the attacker.
- *Data Gathering:* This is the most hands-off approach to social engineering, as it rarely involves having to directly interact with the

target. Every cyber-attack, whether carried out through the network or socially, requires some reconnaissance, some information, as a basis for decision-making. In the context of social engineering this dimension aims to gather as much information as possible to facilitate the primary attack which will lead to the loss, theft, or unavailability of data.

It is sometimes the case that in data gathering, an attacker may accomplish the goal without having to go through the other two dimensions. For example; where an attacker searches through an unsecured waste paper bin and finds passwords for the system written on a piece of paper. This dimension illustrates the importance of physical security, making sure that any sensitive data that is on paper, is shredded and is correctly disposed.

Phishing

Phishing is another form of cyber-attack that does not rely on software to be effective; it is the primary approach used in social engineering attacks. Phishing is most commonly seen in email links that are crafted to be clicked on; which then direct the victim to attachments or weblinks which infiltrate the system and perform malicious functions. Unlike malware, phishing emails bait victims into giving up personal information, such as passwords and other key details, through various tools and techniques, and it does not rely purely on code. Social engineering and social media networks play an important role in successful phishing attacks; these channels are used to directly message a victim to fool them into giving up security details that should otherwise remain secret. Phishing can be incredibly successful if done well, with some research showing a 90% success rate with well-designed phishing websites fooling even the most cyber conscious individuals.

Phishing may be categorised to distinguish the different ways an attacker can find gaps in security systems. The most common are the four vectors, the 4V's, listed below.

- ***Phishing:*** *The traditional form of phishing is through email as described above, an example is given in figure 4.2. Email phishing and website phishing are particularly effective when targeting an organisation's employees, as most users are unaware of what to look for when identifying browser-based cues; status bars, address bars, and other security indicators. The attackers attempt to bait the victim into clicking a link or entering personal information into a phishing website.*

- ***Vishing:*** *Short for "Voice Phishing" is the use of social engineering over a phone network. Attackers may also use VoIP (Voice over IP), such as Skype, where they may spoof the caller ID into something that appears as a trusted landline number. The primary purpose for Vishing is to steal credit card information and other valuable bank details.*
- ***Smishing:*** *Is a form of phishing using SMS text as the medium of delivery. The attackers use the texting abilities on a mobile phone to bait victims into divulging personal details. An example of smishing, is a text offering a gift or monetary prize that can be claimed by following the instructions, which will eventual lead the victim to a phishing website or it might install a form of malware onto the phone.*
- ***Impersonation:*** *Is the most common form of phishing and a degree of impersonation is present in all types of phishing (more detail given in Social Engineering). Impersonation is the primary tool for an attacker, and they will often masquerade as a person with authority and use social engineering techniques to accomplish their goals.*

Figure 4.2: Example of a phishing email

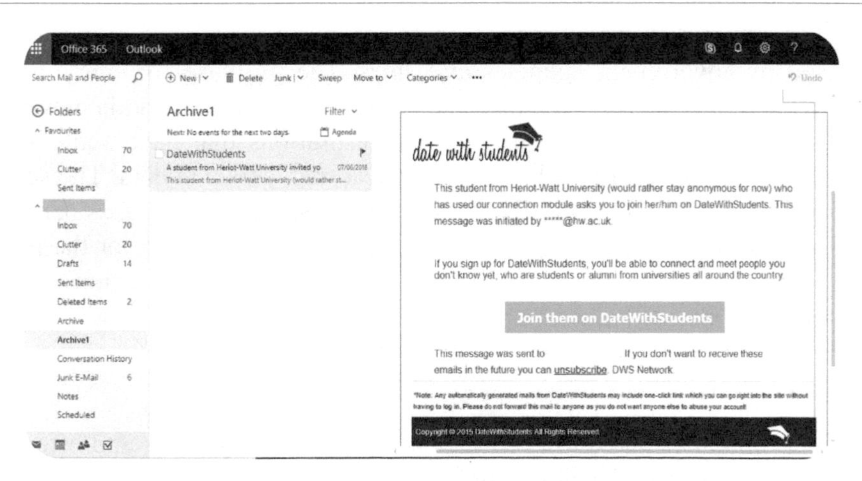

This is an example of a phishing email. The date with students, a fake organisation, sent out automated messages to various ac.uk accounts (which are email accounts for UK based universities). The email attempted to dupe victims into clicking the link by suggesting another student would like to meet them. The link would then gather personal data details on the subjects affected.

Countermeasures

This section will cover cyber-attack countermeasures, primarily by analysing the ways in which cyber security professionals employ hardware and software tools to better protect the information system in question. Within the GDPR, encryption along with pseudonymisation, is the main focus in achieving compliance under the technical safeguards defined in the regulation. The most effective way to secure data, and avoid any risk associated with data breaches, is to create the correct organisational mentality. This mentality is engendered through high-level support from management and in the implementation of relevant personnel training.

Cryptography is a science that deals with the encryption and decryption of messages, known as plaintext. Before the information age, cryptography was used by government agencies (and probably some criminal organisations) for the secure sharing of information. During the second world war, the Enigma machine was one of the most widely used encryption devices used by the Nazi regime to secretly send messages and battle tactics between battalions and other involved parties. It relied on scrambling the plaintext message, into what is known as ciphertext, with a secret "key" or code, which was then decrypted on the other end using the same key.

Encryption has evolved since the second world war and is no longer just used by government agencies and spies. The use case for encryption has moved to the internet as a means of protecting user data and is now widely used by almost all organisations and institutions that have some kind of online presence, especially those that deal with personal data. The scope of cryptography could be a whole textbook in itself (which it is, as part of this series of books in information science and technology), so for the purpose of this section, encryption will be analysed solely as a function of GDPR compliance.

Encryption

Encryption has had a long history in government agencies and often high-grade encryption development has been blocked by these same agencies. It was only in 1973 that the National Bureau of Standards, now known as the National Institute of Standards and Technology; NIST, called for a new method of creating a cipher with an encryption algorithm and a key.

By 1976, this new cipher method became the Data Encryption Standard (DES) cipher. The strength of the DES cipher came from the 64-bit key utilised in the cipher; its weakness was that the two people using the DES cipher would have to share their private key with each other. This created a problem in that a third person could listen in and discover what the private keys were[3].

In 1976 the issue was investigated by Whitfield Diffie, Martin Helman, and Ralph Merkle who proposed a method of using public keys; keys that were publicly known or available. The public key would be used to encrypt the data, whilst the private key only known by the recipient of the message would be used to decrypt the ciphertext.

Put simply, the idea is that, Bob gives out unlocked padlocks (public key) to everyone who wants to communicate with him, then Alice writes a message (to Bob) that is secured by Bob's padlock (public key), which can only be unlocked by Bob who is the holder of the padlock key (private key). This method is known as asymmetric encryption; where only one party can decrypt the message.

Types of Encryption

There are three main types of encryption that used today, the first is symmetric encryption, the second mentioned above is asymmetric encryption, and lastly One-way hash functions. The differences between the two stems from the kind of key used in the encoding of messages.

Asymmetric encryption as exemplified above:

- Uses two keys; a public and a private key.
- The sender receives the recipient's public key.
- The sender then uses a combination of the recipient's public key and their private key.
- The message is then delivered to the relevant person.
- The recipient then receives the senders public key.
- The message is then decrypted by the recipient using a combination of their private key and the senders public key.

[3]In cryptography the two people exchanging encoded messages are often referred to as Bob and Alice, with the third malicious player called Eve. From now on these three names will be used to refer to the exchange in messages.

There are a few well known, and widely used methods of *asymmetric* encryption, namely the Diffie-Helman method (developers of the public key), the RSA method; developed by Ronald L. Rivest, Adi Shamir, and Leonard M. Adleman, which uses the factorisation of prime numbers as the basis for the encryption (the idea being that is it very difficult to find the factor of a prime number using only other prime numbers, also considering the computation power required when the prime numbers are in the 1000's of digits), and lastly ElGamal which uses discrete logarithms in its encoding algorithm. Asymmetric encryption is considered to be more robust than symmetric encryption as it eliminates some of the weaknesses of symmetric encryption. However, with Moore's law[4] it is becoming increasingly difficult for encryption methods to keep up with increasing computational power and its use in cracking encryptions.

The second type of encryption is known as *symmetric* encryption, where both the sender and receiver have the same private key; it uses one key pair that both encrypts and decrypts the message. The weakness of symmetric key encryption is that data can be lost if an attacker were to intercept the private key in some way, leaving symmetric encryption open to brute force attack; an attack where trial and error (guessing) software is used to crack encrypted data.

The final method of encryption is *hashing*, or one-way hashing functions. Hashing is used to hide the contents of a message or to check the integrity of data. In a one-way hash function, it should not be mathematically possible to reverse the cipher (hash value) back to the original data. Unfortunately, it can be broken by knowing how the data was mapped to the hash value, or by performing a brute force analysis on the stored hash value. The one-way hash function is typically used in authentication applications, such as generating a hash value for a message, and to store ciphered versions of passwords. The method of uncovering the mapping between the hashed values and the original data is called a *rainbow table attack,* while a brute force analysis is known as a *dictionary-type attack.*

Below are a couple of hashing methods used today:

- *Base-64 Encoding:* This is used in electronic mail and is typically used to change binary data into a standard 7-bit ASCII form. It takes 6-bit characters, at a time, and converts them to a printable character.

[4]This law observes that the number of transistors in an integrated system doubles about every two years (computational power). What this means for cryptography is that key space must keep up with the trend, such as 64-bit keys developed in 1970's are useless at protecting data as the amount of computational power today could crack a 64-bit key in about a day.

- *UNIX Password Hashing:* This is used in the *passwd* file which contains hashed versions of passwords. It is a one-way function, so that it is typically not possible to guess the password from the hashed code. If the hashed code for the given word is known, an attacker can determine the password and bypass the encryption. Weak passwords are subject to dictionary-attacks, where offline programs can be used to search through a known dictionary of common words and then match the hashed code against the one in the password file.
- *MD5:* This type of hashing is used in several encryption and authentication methods. It produces a 32 hexadecimal character output (128 bits), which can also be converted into a text format.
- *SHA (Secret Hash Algorithm):* This is an enhanced message hash, which produces a 40 hexadecimal character output (160-bits). It will thus produce a 40 hexadecimal character signature for any message from 1 to 2,305,843,009,213,693,952 characters. At present it is computationally difficult to produce two messages with the same hashed value.

Function of cryptography

The primary function of cryptography, and by extension encryption, is similar to the sprit of the GDPR, in that it encourages a platform in which two or more parties can communicate without the fear of other entities, malicious or otherwise, might interfere with their basic rights. Cryptography, or key-based cryptography (encryption), function as:

- Integrity Check: Making sure that the message has not been tampered with by non-legitimate sources.
- Providing Authentication: Verifying the sender's identity, with the openness of the internet today, is proving to be challenging for individuals.

Pseudonymisation

Pseudonymisation is a process of data management; it strips away certain parts of a data set that can be used to identify the data subjects. The stripped parts of the data set are then replaced with non-identifiers, or pseudonyms, and the data set can then only be recovered and matched back to the data subject with the correct encryption key. Pseudonymisation allows the data processor or controller to process the data without jeopardising the rights of the data subject, it is an extension of the encryption that is cited within the regulation.

When pseudonymising data, it is up to the organisation to choose which fields to pseudonymise. Data fields which require this attention are those with strong personal identifiers; names, addresses, health records, etc. It may not be necessary to anonymise other data fields that would be openly available elsewhere, such as public records.

Chapter Review

Summary of Key Points

1) 3 P's: Privacy, Protection, Process.
2) Types of cyber-attacks: software, social.
3) Types of Malware.
4) Attacks using social engineering.
5) Countermeasures.
6) 3 types of encryption.

- Have you received a phishing email?
- How would you explain a cyber-attack to a non-technical person?
- What staff training would you suggest to counter-act social engineering?
- How would you explain asymmetric encryption to senior management?

References

Buchanan, W., Li, S., & Asif, R. (2017). Lightweight cryptography methods. *Journal of Cyber Security Technology, 1*(3–4), 187–201.

Cialdini, R. (2001). *Influence: Science and practice (4th ed.).*

Cialdini, R., & Goldstein, N. (2004). Social Influence: Compliance and Conformity. *Annual Review of Psychology, 55*(1), 591–621.

Cialdini, R., & Trost, M. (1998). Social influence: Social norms, conformity and compliance. *The Handbook of Social Psychology, Vol. 2*, 151–192.

Courter, B. (2010). CEOs Vulnerable to Scammers Who Target Whales. *Credit Union Times*, pp. 8–8.

Dhamija, R., Tygar, J., & Hearst, M. (2006). Why phishing works. *Proceedings of the SIGCHI conference on Human Factors in computing systems – CHI '06*, (p. 581).

Diffie, W., Diffie, W., & Hellman, M. (1976). New Directions in Cryptography. *IEEE Transactions on Information Theory, 22*(6), 644–654.

Green, A. (2017). Ransomware and the GDPR. *Network Security, 2017*(3), 18–19.

Intelligence, C. (2017). Cyber Security: Unstructured Data and the Threats You Cannot See. *BIIA blog*.

Mouha, N. (2015). The Design Space of Lightweight Cryptography. *NIST Lightweight Cryptography Workshop 2015*, 1–19.

National Cyber Security Center. (2018). *NCSC Incident Response*. Retrieved from NCSC: https://www.ncsc.govt.nz/incidents/

National Cyber Security Centre. (2018). *Cyber Essentials :: Requirements for IT Infrastructure*. Retrieved from Cyber Essentials: https://www.cyberessentials.ncsc.gov.uk/requirements-for-it-infrastructure.html

Piper, A. (2013). Trapping Hackers. *Risk Management (00355593), 60*(4), 8–10.

Schneier, B. (1998). *Security pitfalls in cryptography*. Retrieved from Schneier on Security.

Schneier, B. (2008). Schneier on Security. *Wwwschneiercom*, 336.

Schneier, B. (2008). The psychology of security. *Lecture Notes in Computer Science (including subseries Lecture Notes in Artificial Intelligence and Lecture Notes in Bioinformatics), 5023 LNCS*, pp. 50–79.

Tetri, P., & Vuorinen, J. (2013). Dissecting social engineering. *Behaviour and Information Technology, 32*(10), 1014–1023.

Willems, G., Holz, T., & Freiling, F. (2007). Toward automated dynamic malware analysis using CWSandbox. *IEEE Security and Privacy, 5(2)*, 32–39.

Wu, L., Du, X., & Wu, J. (2016). Effective Defense Schemes for Phishing Attacks on Mobile Computing Platforms. *IEEE Transactions on Vehicular Technology, 65*(8), 6678–6691.

2

Part Two: Preparatory Steps

Chapters Covered

Section Overview

This part of the book explains the preparatory steps and requirements that need to be considered before taking action. It also details the why and how of data protection by design and default; the reasoning behind the formulation of policies, their structure and the subsequent development of procedures designed to ensure adherence to those policies. General frameworks used in the creation of these documents will be outlined. The final chapter explains the acronyms; DPO, DPIA, and DSAR, and how and where they're relevant.

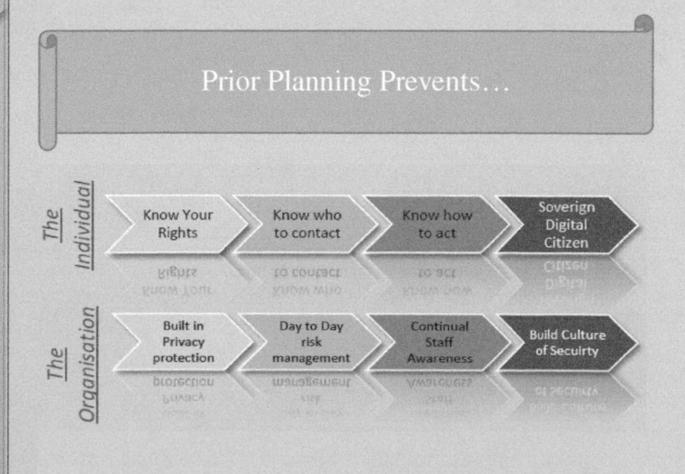

Chapter 5: Data Protection by Design and Default

At a Glance:

- 7 foundational principles of data protection by design
- CIA principle of data
- Organisational measures: building a culture of security
- Physical and computer security
- The OWASP

Learning Objectives: *Students should be able to…*

- Identify the key principles of data protection by design and default.
- Understand the people problem in cyber security.
- Explain the CIA principle of data security.
- Identify Organisational and technical measures of data protection.

Key Terms

1) Privacy by Design
2) Confidentiality, Integrity, Availability
3) Physical Security
4) Computer Security
5) Intrusion Detection System
6) Firewall

Introduction

1. *Taking into account the state of the art, the cost of implementation and the nature, scope, context, and purposes of processing as well as the risks of varying likelihood and severity for rights and freedoms of natural persons posed by the processing, the controller shall, both at the time of the determination of the means for processing and at the time of the processing itself, implement appropriate technical and organizational measures, such as pseudonymization, which are designed to implement data-protection principles, such as data minimisation, in an effective manner and to integrate the necessary safeguards into the processing in order to meet the requirement of this Regulation and protect the right of data subjects.*
2. *The controller shall implement appropriate technical and organizational measures for ensuring that, by default, only personal data which are necessary for each specific purpose of the processing are processed. That obligation applier to the amount of personal data collected, the extent of their processing, the period of their storage and their accessibility. In particular, such measures shall ensure that by default personal data are not made accessible without the individual's intervention to an indefinite number of natural persons.*
3. *An approved certification mechanism pursuant to Article 42 may be used as an element to demonstrate compliance with the requirements set out in paragraphs 1 and 2 of this Article.*

Article 25

Data Protection is a Program; not a Project

There is a saying that the only certainties in life are Death and Taxes; but when it comes to the certainties of organisational practices, it is Taxes and Data protection! In the same way that organisations must meet their financial obligations as an ongoing responsibility; so too must the obligations of data protection be met. A handy mantra might be: Every day, in every way, the security of personal data must be enhanced and ensured.

There is a broad requirement under the GDPR to provide data protection by design and default. The protection of personal data is the foundation of the GDPR, and this requirement establishes the obligation of the controller to implement measures to ensure this protection. The measures are expected to be applied in both the technical and organisational aspects of the enterprise, and in practise this means that consideration must be given to data protection before, during, and after, the implementation of data protection measures. This requirement is easiest to achieve from the beginning of the systems life-cycle, rather than trying to integrate it on an *ad hoc* basis.

Data protection by design and default must become an integral and non-negotiable part of the development and implementation of every organisational system. The requirement within the GDPR emphasises that this is both a technical and an organisational obligation. This means organisations need to cultivate a culture of data protection awareness. Organisational culture is broadly defined as the shared assumptions which define acceptable behaviour within a given organisation. Therefore, the assumptions about how data protection is implemented within the organisation may need to be addressed. Organisation wide training is an important tool in culture change, and this must include training of senior management. Policy creation and review requires the engagement of senior management before procedural implementation can begin, and this highlights the need for up-to-date training.

What is Privacy?

The right that determines the non-intervention of secret surveillance and the protection of an individual's information. It is split into 4 categories[1]:

1) *Physical*: An imposition whereby another individual is restricted from experiencing an individual or a situation.
2) *Decisional*: The imposition of a restriction that is exclusive to an entity.
3) *Informational*: The prevention of searching for unknown information.
4) *Dispositional*: The prevention of attempts made to get to know the state of mind of an individual.

[1] Black's Law Dictionary 2nd ed.

As it applies to data, the Organisation for Economic Development (OECD) defines privacy as:

> *"It is the status accorded to data which has been agreed upon between the person or organisation furnishing the data and the organisation receiving it and which describes the degree of protection which will be provided."*

Informational privacy includes the freedom from intrusion, into an individual's data, by those seeking to gather information. It also includes the ability to choose how, and to what degree, personal data can be shared, or withheld, from others. Privacy is the right to exercise freedom in the control of one's data flow; that is in the collection, processing, and sharing of personal data.

Privacy and Protection by Design and Default

Privacy by design is not a new concept. In the late 1990's the Information and Privacy Commissioner of Ontario, Canada; Dr Ann Cavoukian, developed the *Privacy by Design* (PbD) approach as a practical tool for improving data privacy provision and protection. The prevailing attitude at the time was one of "either/or", or as Cavoukian termed it; the Zero-sum model. This model suggests that privacy must necessarily be sacrificed to achieve certain desired outcomes; PbD presented an alternate approach; the Positive-sum model. This model allows functionalities, for example; privacy and security, to create positive gain in both.

Privacy by Design (PbD) was adopted as an international standard in 2010 and is incorporated into the GDPR. As a tool for best practice it is a proactive method for privacy protection, employing 7 Foundational Principles:

1. Proactive not Reactive; Preventative not Remedial

The Privacy by Design (PbD) approach is characterized by proactive measures; designed at the beginning stages of new technologies or organisational processes. PbD is a preventative tool to be used before problems arise.

2. Privacy as the Default Setting

Privacy is embedded in the system; as the default setting. Implementing PbD allows the maximum degree of privacy to be provided, through automatically protecting personal data in any given IT system or organisational process. Inaction by the individual (data subject) provides maximum privacy protection as a default setting.

3. Privacy Embedded into Design

It is embedded into the core functionality of the organisation, ultimately becoming part of the culture. This engenders trust in the organisation's ability to protect privacy whilst preserving functionality.

4. Full Functionality — Positive-Sum, not Zero-Sum

Designing incremental functional improvements into the system; where all functions involved improve, rather than improving one at the expense of another.

5. End-to-End Security — Full Lifecycle Protection

Critical to PbD, protection must be provided throughout the full lifecycle of the data involved, to ensure the secure retention, and secure destruction, of the data at the end of the process.

6. Visibility and Transparency — Keep it Open

Stakeholders are assured of the organisation's integrity regarding the stated promises and objectives, subject to independent verification. This is provided through visible and transparent component parts and operations; allowing users and providers to, trust and verify.

7. Respect for User Privacy — Keep it User-Centric

Systems architects and operators are required to keep the interests of the individual paramount. By providing strong privacy defaults, appropriate notices, and empowering user-friendly options; the key point is to keep it user-centric. Remember that the data belongs to the individual; the organisation has custody and some control over it; but not the ownership.

At its core, privacy and protection by design puts the rights of the individual into the design process alongside the goals of the organisation. Data subjects, are given greater control over how, where, when, and with whom their data is processed, through easy to use tools which have been designed into the system for this specific purpose. These tools should help data subjects to manage the processing of their data.

Building on the principles of PbD, the default setting of all systems dealing with personal data must provide maximum security and privacy. Protection is incorporated to the degree that the privacy of an individual's data is intact where there is NO action by that individual. The accountability for this protection lies with the Data Controller, and they have a responsibility to show their compliance. Depending on the size of the organisation and the scope of influence in their field, the level of accountability may increase along with the level of scrutiny.

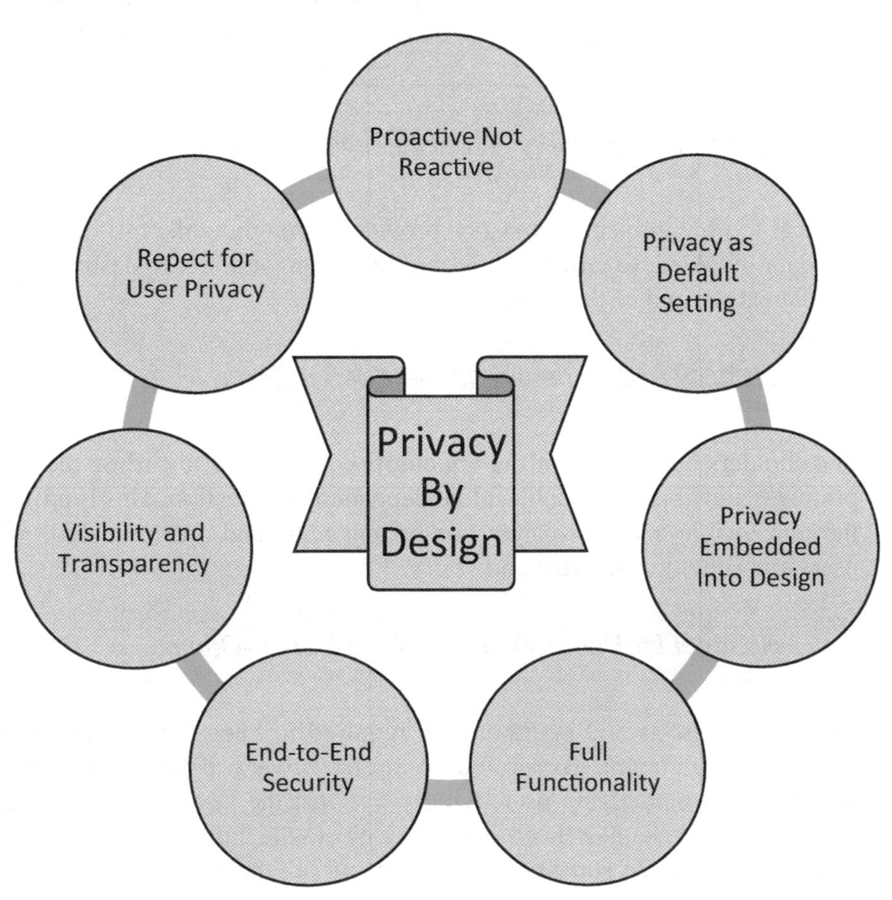

The Security Principle: Appropriate Technical and Organisational Measures

What are the technical and organisational measures referred to within the GDPR? They refer to the "security principle"; that is the assessment and implementation of the appropriate level of security to protect the personal data held. This protection must cover accidental, as well as deliberate, interference or compromise and must be tailored to the specific circumstances of the organisation.

The security principle is addressed in part through Information Security (InfoSec). This is the field in which the integrity, confidentiality, and availability of information is assured through information security management systems. These systems include physical material information (paper etc.) and digital information.

CIA Principles

Confidentiality

Principle #1

- The objective of this principle is to protect information that is classified as confidential.
- Promoting privacy.

Integrity

Principle #2

- The system that processes data is complete, timely, and authorized.
- Ties-in with the 4th and 6th principles of the GDPR.

Availability

Principle #3

- System is available for operation and use as agreed.
- Ties-in with the Rigths of the Natural Persons under the GDPR.

Risk assessment is emphasised as an integral aspect of the security principle in the GDPR, as specified in: Article 32(1)

Taking into account the state of the art, the costs of implementation and the nature, scope, context and purposes of processing as well as the risk of varying likelihood and severity for the rights and freedoms of natural persons, the controller and the processor shall implement appropriate technical and organisational measures to ensure a level of security appropriate to the risk.

Article 32(1)

Therefore, when deciding on the measures to be used in applying the security principle to the protection of personal data, the risks posed to the individual arising from the specific circumstances of the processing, must be assessed. These risks include lack of access and availability to personal data, by the data subject. Risk assessment standards and methods are discussed in Chapter 9.

Organisational: A Corporate Culture of Data Protection

Understanding how, and by whom, data is accessed and used within an organisation is critical to the protection of that data. Organisational measures are those which relate to the systems environment and to the way in which people interact with it and each other. A risk assessment is an example of an organisational measure, as is the creation of an information security policy. This includes staff, departments, vendors, clients, and other third parties. When we speak about organisations we must consider the culture, and the promotion of a culture of data protection begins with accountability. A clear and well communicated policy for information security will outline the goals and responsibilities, and will define where that responsibility lies within the organisation.

The size and structure of the organisation will determine, in part, the nature and scope of the security policy. However, any organisation regardless of size can use its information security policy as part of the documentation to show compliance. Policies are created at the senior management level and require a holistic approach to assess the inter-relatedness of the various departments and processes within the organisation. Business Continuity Management, as part of Risk Management, provides useful frameworks prior to the preparation of policies and procedures and is discussed in Chapter 9.

Third parties are also included in this requirement. Addenda to the GDPR extends the concepts of data protection by design to other organisations, although it does not place a requirement on them to comply – that remains with the controller.

Recital 78 says:
When developing, designing, selecting and using applications, services and products that are based on the processing of personal data or process personal data to fulfil their task, producers of the products, services and applications should be encouraged to take into account the right to data protection when developing and designing such products, services and applications and, with regard to the state of the art, to make sure that controllers and processors are able to fulfil their data protection obligations.

Staff Awareness of Security

All the security tech in the world will not protect data from a staff member's ignorant actions and remedying that lack of knowledge is the responsibility of the organisation.

A key organisational measure is in the training of staff to be aware of their duties and obligations around data protection. This includes education in the types of risks they may face in their work as it relates to the processing and protection of data. The behaviour of the members within an organisation plays a significant role in the creation of a culture of security. Cyber security may be seen as a "departmental" or IT issue rather than one in which the organisation as a whole, and each of its members, must be engaged in. Therefore, it is important to demonstrate a "top-down" engagement with data protection, and staff at all levels should see and understand the necessity for data protection as modelled by senior management.

Training for end users is often considered as an afterthought to technology-based countermeasures, to the detriment of security provision as a whole. Often training is provided as a "one type serves all" approach and is not developed with the needs of the end user in mind. Unsurprisingly, this achieves little in the way of improved security habits for the trainee and may lead to increased risk exposure for the organisation.

The risk arises from inadequate training that gives rise to a superficial awareness of the actual threats faced by the staff member, and the organisation;

> *"Well, the average computer user is going to pick dancing pigs over security any day. And we can't expect them not to."*
>
> Bruce Schneier, 1999.

This quote from cyber-security expert, Bruce Schneier refers to the habit of many end users, at work and at home, to ignore security pop-up screens warning them of the presence of malicious code. When this happens on a workstation the results could be devastating for the organisation, so why does it still happen? One of the fundamental reasons for the ignoring of security warning messages is lack of knowledge; end users do not understand the basis on which these notices arise.

Added to this is the complex wording of these messages without a clear statement of the benefits gained from following the security advice. This applies to password creation as much as it does to opening malicious emails and highlights the need for targeted and specific training which engages the end user as an affiliate member of the Information Security team.

Tick-box training exercises and posters will not demonstrate a culture of security awareness. Relevant training delivered through the medium most appropriate for the staff member, aimed at increasing understanding, and the development of skills, will produce better results and provides measurable outputs. These outputs can then be used to improve the training and awareness program and included in documentation.

Key Areas for Staff Awareness

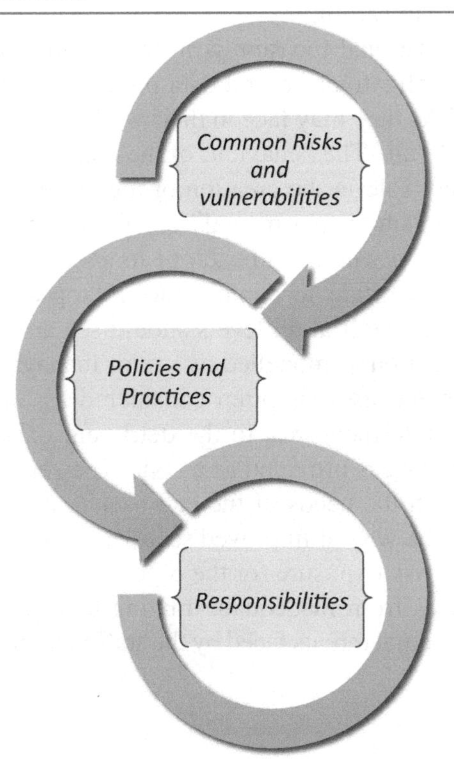

Organisational Responsibility for Security

The protection of data and information within an organisation is a people problem; that is, one of human interaction. This interaction involves the relationship to data protection and security as concepts, and the relationship between members of the organisation. For security to be effective, there must be organisation wide engagement; just as when someone leaves the fire exit door open, or leaves their workstation unlocked, these actions can have wide-reaching consequences.

Organisational security addresses the level of access staff have to the data within the organisation; analysing why that access is necessary, or not, and how that access is implemented. Fundamentally, it seeks to guide and control organisational behaviour as it relates to information security. One key aspect is to identify who is responsible for data security, and which levels of access and accountability are appropriate to those individuals.

Technical Measures

At first blush, technical measures may be assumed to refer solely to computer security; the data stored within networks and computers however, the protection of *physical* assets is also a technical measure. Physical hardware and documents containing personal data must be kept secure and when no longer of use they must be disposed of correctly. They should be destroyed in a secure manner, so that any data contained within them is irretrievable. Within the computer security, IT context, the security of key factors must be addressed; systems, data, devices, and online activity. It may be necessary to consult specialists external to the organisation for advice on areas in which the organisation has limited knowledge.

Physical Security

Physical security is the foundation of data security. All the software protection and security training in the budget will not help if unauthorised physical access to computers is made easy. Likewise protecting staff, networks, software, hardware, and the data itself, from physical interference is the first step in providing physical security. This interference may be in the form of non-malicious external events; such as natural disasters, flood, fire (including arson), or organised attacks; like theft, manipulation (of people and things), or broader acts of revenge or terrorism.

Unlike technical cyber-attacks, physical threats can cause significant damage with little to no technical expertise on the part of the attacker. This makes the protection of physical assets high-risk, in that they may be compromised by a greater number of factors or actors.

Case Study: Cayla Doll

In 2017, the German government banned the My Friend Cayla doll, after it was proven to be a surveillance device.

The "doll" connects to the internet via WIFI and has inbuilt speech recognition and cameras, all of which is unsecured. Hackers could use the Bluetooth connection to listen and speak to the child through the device, and the cameras could be accessed remotely to spy on the child.

The IoT enabled device was determined to be an, "illegal espionage apparatus", since it was identified as a surveillance device disguised as a toy; and in Germany this is illegal.

The device is manufactured by Genesis Toys in the USA and is still on sale around the world.

The three areas of physical security are:

Access control: the blocking of physical access to secured sites; including locked doors and cupboards, security fences, staff access cards, visitor management, and anti-fire and flooding systems.

Surveillance: the monitoring of physical locations visually and digitally, heat and smoke sensors, and intruder detection systems.

Testing: regular review using Business Continuity Management (BCM), of policies and procedures on disaster recovery.

The Internet of Things (IoT)

The Internet of Things (IoT), refers to the network of physical objects (things) which are able to connect to the Internet, wirelessly. These interknitted-things are embedded with technology that enables the collection, storage, and transmission of data. The objects cover a wide range from: medical devices, toys, phones, buildings, vehicles, fridges, and many, many more. As the implementation of the IoT has occurred rapidly, the vulnerabilities it creates have been dealt with as an afterthought. Cyber-attacks were previously contained within the digital world, now through the IoT, the attack may be physical.

Internet connected wireless devices pose a specific risk in the implementation of security protocols, as they can remain connected to organisational systems when outside physical security boundaries. Physical measures to enhance device protection and security are a basic first step:

1) Keeping track of the device is important through device location technology.
2) Tamper-proof locks and I.D. tags
3) Motion sensors and tracking signals

Hardware Security

As with IoT devices, hardware also has security risks which must be addressed as part of a complete security approach. Vulnerabilities may be found anywhere in the system, so it is necessary to identify where these occur and how they may be remedied, for example:

External drives:

- Open USB ports allow access for dongles with malicious software.
- Disable external drives where relevant.
- Use USB dongle keys access.

Computer case security:

- Lock cases to prevent the hard drive from being removed.
- Install drive locks to encrypt hard drives.
- Use intrusion detection program to alert user to interference.

Workstation security:

- Unplug and lock away unused computers.
- Lock offices when users are away.
- Screen computers from public view.
- Automatic log-out if inactive for set period.
- Smart card access only. Publicly accessible areas, such as the reception area are particularly vulnerable and require a tailored approach to enhance physical security whilst still fulfilling organisational needs.

Portable Devices:

- Lock away portable devices when not in use.
- Encryption software on every device.
- 2-factor authentication as default.

Printers:

- On-board memory can be accessed without authorisation and documents released.
- Use a secure printer with encryption!
- Change the default passwords.
- Employ smart card access.
- Lock printer room when not in use.
- Institute "shred it all" procedure.
- Provide a shredder at every printer.
- Install camera surveillance.

Server room:

- Keep it locked always.
- Smart card and biometric access only.
- Cage servers and secure to floor.
- Disconnect unused network connections.
- Employ Internal and external surveillance cameras.
- Install heat and motion detectors and fire suppression system.
- Store data backups off-site or in separate secure room.
- Store vulnerable or critical hardware in cages in secure room.
- Ensure adequate ventilation of the room.

Computer Security: Design

Computer security by design is an approach to software development which assumes that insecure or malicious practices will occur. Thus, software is designed from the very outset with the intention to minimise security vulnerabilities. Some of the main techniques used in a security by design approaches are outlined below.

Principle of least privilege or authority:

Limited access based on the minimum requirement needed by users to perform tasks. This is akin to the lowest security level clearance. Least privilege principles can also be applied to processes, systems, apps, devices, and hardware; limiting access to minimal parts of the system.

Automated Theorem Proving (ATP):

Uses automated reasoning to prove the correct functioning of software sub-systems.

Audit trails:
Tracks system activity and used to analyse the method and extent of any attacks or breaches which have occurred. Must be stored remotely in an "append only" mode.

Code review and unit testing:
Used to identify vulnerabilities through errors or malware in program code, may be performed personally by developers, or using an automated review system. Unit testing is performed to check functionality and eliminate errors during construction.

Defence in depth:
Principle of layered security where many layers of security measures are designed into the system which lead to increased security overall. If one mechanism is compromised, other layers of security which work to enhance the security of that mechanism, will provide protection.

Fail Secure:
Design that incorporates a fail state which automatically defaults to a maximum-security setting. Secure systems should be designed so that an insecure state can only be created by those authorised to do so.

Full disclosure:
When vulnerabilities are found ensure the exposure is minimised by immediately disclosing bugs or weaknesses.

Computer Security: Measures

There are three processes used in the quest to achieve computer security, and these processes revolve around analysis of threats. Security measures aim to prevent, detect, and respond to threats. They are implemented through organisational means; such as policies and procedures, and technical means; such as system components. Some fundamental measures are as follows:

Intrusion Detection System (IDS):
These detect in-progress network attacks by monitoring network traffic flows for anomalous activity. Passive IDS detect suspicious activity and sends an alert to the user or administrator. Reactive IDS also provide for active response to the threat, usually by blocking further exchange from the source. Network Intrusion Detection Systems (NIDS) are placed within the network

and monitor all traffic, in and out of the system. Host Intrusion Detection Systems (HIDS) provide the same function on individual devices or hosts.

User Access Controls (UAC):
Managing user accounts to prevent unauthorised access. Applying the Principle of Least Privilege, accounts with special access privileges should be extended to authorised staff only. Accounts with Administrative privileges should not be used for general work or connect to the internet. This reduces the exposure to external attackers. Software should be installed from official vendor sites only to minimise exposure to malware.

Firewalls:
Hardware and software firewalls are an essential component, when properly configured. Providing a shield between external and internal networks; blocking attacks. Next-generation firewalls (NGFW) and Threat-focused NGFW's, provide additional capabilities in the form of intrusion prevention, ability to address active security threats, identify at-risk assets, and shorten response-cycle interval.

What is Security?

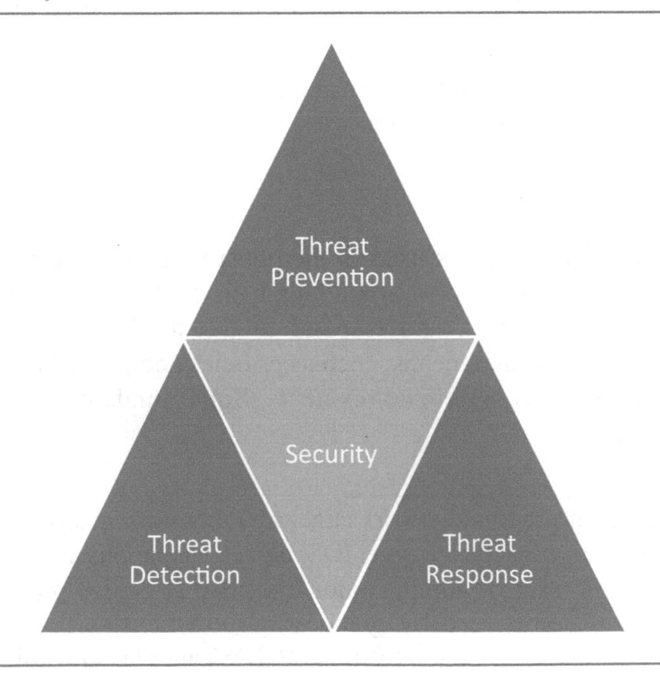

Open Web Application Security Project (OWASP)

"Software powers the world, but insecure software threatens safety, trust, and economic growth."

The Open Web Application Security Project is a worldwide, free, and open source software security project and community. It is an open community, with over 46,000 members, which works to encourage and enable organisations in every aspect of software security. Tools are developed to engender trust in application security. It is a not for profit organisation which focusses on improving software security, worldwide. Primarily dealing with Application security, its main mission is to make software security visible and to develop "best practice" tools and methodologies.

The OWASP Foundation is an independent, not-for-profit organisation (NFP), funded by individuals and organisations who believe open-platform security methods lead to the development of trusted applications. As an NFP, the tools, documents, and applications developed through OWASP are offered free of charge. Engendering trust is a key goal and is underpinned by the four core values of the project. These are: (Code of Ethics found in the appendix)

- *Open*: Transparency is key.
- *Innovation*: Solutions to security challenges.
- *Global*: All inclusive.
- *Integrity*: Vendor neutral.

As a community with global reach, the *OWASP Top 10* report on the most critical security risks, is highly respected within the software industry. Companies, such as Oracle, respond to the vulnerabilities listed in the *OWASP Top 10*[2], which is updated periodically and published under the Creative Commons license.

The current *OWASP Top 10*, 2017:

1) *Injection*: Injection flaws occur when malicious data is sent to an interpreter.
2) *Broken Authentication*: Attackers can compromise a system with access to few accounts.

---

[2] For example: *Security in Oracle ADF: Addressing the OWASP Top 10 Security Vulnerabilities.*

3) *Sensitive Data Exposure*: Clear-text data (unencrypted data) or keys are stolen.
4) *XML External Entities*: Can extract data, scan systems, and execute attacks.
5) *Broken Access Control*: Attackers are able to act as Administrators or users.
6) *Security Misconfiguration*: Unpatched or badly configured systems create exploitable flaws.
7) *Cross-site scripting (XSS)*: Malicious script is allowed to remain on a new web page.
8) *Insecure De-serialisation*: Allows one of the most serious attacks; remote code execution.
9) *Using Components with known Vulnerabilities*: Software modules which are unpatched, old, or un-monitored are easily exploited.
10) *Insufficient Logging and Monitoring*: Lack of monitoring and recording of login activity, and slow or ineffectual response to incidents.

OWASP recommend the use of repeatable security processes through the implementation of standard security controls. This should be designed-in to the development process using a secure development lifecycle such as the; OWASP Software Assurance Maturity Model (SAMM).

Assessing Information Assets: Value and Risk

In order to design systems with data protection built in, it is necessary to understand and classify the information according to the level of risk posed to the data subject if a breach occurs. Consistent and logical labelling of personal data will facilitate retrieval, rectification, and erasure[3]. Data mapping, as discussed earlier (Chapter 3), provides an overview of where the data flows in and out of the system, and where it is at rest (held), and who utilises that data. It is then necessary to classify the data into risk types; the standard for applying the appropriate level of classification is detailed in the ISO 27001 control objective, '*Information Classification*', and is explained in Chapter 9. We will briefly outline some of the considerations to be aware of when classifying and labelling information.

[3] Refer to Chapter 1; Rights of the Data Subject.

Information Classification and Labelling

Classification:

The process of assessing the value of the data being processed and classifying it according to its level of risk. This classification facilitates the development of policies and procedures[4] relevant to the subsequent handling of each classification type. As a starting point, policies for information classification must address what classification labels and descriptions are to be used in a particular context. Staff should readily understand which level of protection should be applied, or not, to any information they interact with within the organisation. The naming of the classifications is not standardised, however the term 'high risk' is used within the GDPR and this presents an opportunity to name classifications as follows:

- *Public:* open information which does not require protection.
- *Internal Only:* organisational data, not for public distribution.
- *Protected:* Information which is protected by law, such as personal data.
- *Restricted:* Confidential information which may be accessed by a limited number of people, in the performance of their duties.
- *Confidential:* Information that is not widely known outside of the individual, or organisation, to whom it relates. This includes trade secrets or medical records.
- *Secret/Top secret:* Information which if released, could endanger life, society, or the greater functioning of the organisation.

Using the above classifications, it is then possible to apply risk classifications in finer detail. As the GDPR emphasises the risk, from the point of view of the consequences to the data subject, the following categories must be assigned:

Low Risk: data safe for general distribution
High Risk: sensitive data
Very High Risk: data which is very sensitive

[4] Creation and implementation of policies and procedures is discussed in Chapter 6.

Special Category Data: Sensitive and Very Sensitive Personal Data

The GDPR refers to sensitive personal data as "special categories of personal data", and the conditions for processing this type of data are outlined in Article 9(2), found in the appendix due to its length. The lawful basis for processing this data must be stated *and* the specific conditions outlined in Article 9 must also be met. Special data includes information regarding an individual's:

- *health*
- *genetics*
- *biometrics (where used for ID purposes)*
- *religion*
- *sex life*
- *sexual orientation*
- *race*
- *ethnic origin*
- *politics*
- *trade union membership*

Criminal Offence Data

Criminal offence data includes the type of data about criminal allegations, proceedings or convictions that would have been classified as sensitive personal data under previous legislation[5]. However, it may be broader than this, as Article 10 specifically extends the nature of criminal defence data to personal data linked to related security measures. Personal data relating to criminal convictions and offences also requires extra safeguards, similar to Special category data. This category of data can only be processed under specific legal authorisation, or in an official capacity. The lawful basis for processing under Article 6, is also required as well as the requirements in Article 10.

Labelling of Data

Once data has been classified according to the level of risk, it is important that the recipient of that data, or the end-user is made aware of the classification level. This allows the recipient to handle the data correctly, in accordance

[5] The Data Protection Act, 1998.

Processing of personal data relating to criminal convictions and offences or related security measures based on article 6(1) shall be carried out only under the control of official authority or when the processing is authorized by Union or Member State law providing for appropriate safeguards for the rights and freedoms of data subjects. Any comprehensive register of criminal convictions shall be kept only under the control of official authority.

Article 10

with the policy guide lines. In some instances, there may be different classifications of data within one document, and these must be highlighted to ensure their correct protection by the recipient. This may be done through differing colours, font styles, or sidebar notices which serve to draw attention to the nature of the data. The method chosen must be consistent across the organisation and outlined within the policy document. Similarly, multi-national or third-party recipients should understand the classification without the need to translate or interpret the document.

Labelled data may be further compartmentalised by the use of descriptors. These are labels which give added information about the level of classification, so that end-users can clearly understand the classification and how it should be handled. Common descriptors highlight to whom the information should be shared, for example; INTERNAL ONLY, or the type of data contained, such as; RESTRICTED – Medical. The point is to facilitate the safe and appropriate handling of all the data within the organisation, and this requires clear, enforceable, top-down policies and procedures to be implemented within the organisation.

Chapter Review

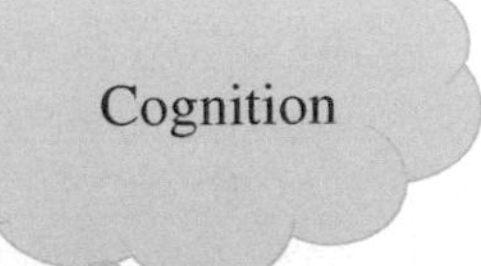

Summary of Key Points

1) 7 foundational principles of Privacy by Design.
2) Security Principles and the CIA, organisational measures and technical measures.
3) Organisational responsibility for security.
4) OWASP.
5) Assessing Information Assets (Information Classification).

- ❖ What are the key steps in designing a well-developed privacy policy?
- ❖ How would you, using the 7 foundational principles, design security into your organisation?
- ❖ What are the 3 elements of data security?
- ❖ What is the processes of information classification? How do special categories of data fit in?

References

Beyer, Richard E.; Brummel, B. (2005). *Implementing Effective Cyber Security Training for End Users of Computer Networks.* Society for Human Resource Management and Society for Industrial and Organizational Psychology.

Black's Law Dictionary. (2018). *What is PRIVACY?* Retrieved from The Law Dictionary: https://thelawdictionary.org/privacy/

Cavoukian, A. (2011). Privacy by Design:The 7 Foundational Principles. 1-2. Toronto: Information & Privacy Commissioner Ontario.

Cavoukian, A. (2012, 12). Privacy by Design. *IEEE TECHNOLOGY AND SOCIETY MAGAZINE*, pp. 18–19.

Cisco. (2018). *What Is a Firewall?* Retrieved from Cisco.com: https://www.cisco.com/c/en/us/products/security/firewalls/what-is-a-firewall.html

Clark, K. (n.d.). Automated Security Classification.

Couvakian, A; Dixon, M. (2013). *Privacy and Security by Design: An Enterprise Architecture Approach.* Information and Privacy Commissioner, Toronto.

Dougherty, Chad., Sayre, Kirk., Seacord, Robert., Svoboda, David., & Togashi, K. (2009). *Secure Design Patterns.* Software Engineering Institute., Pittsburgh.

ENISA. (2016). *Guidelines for SMEs on the security of personal data processing*. Retrieved from Guidelines for SMEs on the security of personal data processing.

Eureka. (2017). *Document Security Under GDPR – a legal perspective.* Retrieved from www.eureka.eu.com: https://eureka.eu.com/gdpr/document-security-gdpr-legal-perspective/

European Data Protection Supervisor. (2018). *Data Protection.* Retrieved from Website: https://edps.europa.eu/data-protection_en

Fernandez, E., & Larrondo-Petrie, M. (2007). Security Patterns and Secure Systems Design. *Proceedings of the 4th LACCEI International Latin American and Caribbean Conference for Engineering and Technology (LACCET'2006)*(June), 233–234.

Herley, C. (2009). So long, and no thanks for the externalities: the rational rejection of security advice by users. *New security paradigms workshop* (pp. 133–144). Oxford: ACM New York.

Information & Privacy Commissioner of Ontario. (2013). Privacy by Design. 1–6. Toronto: Information & Privacy Commissioner of Ontario.

Information Commissioner's Office. (2018). *Data protection by design and default.* Retrieved from Guide to the General Data Protection Regulation (GDPR): https://ico.org.uk/for-organisations/guide-to-the-general-data-protection-regulation-gdpr/accountability-and-governance/data-protection-by-design-and-default/

Information Commissioner's Office. (2018). *Privacy by design.* Retrieved from Guide to data protection: https://ico.org.uk/for-organisations/guide-to-data-protection/privacy-by-design/

Information Commissioner's Office. (2018). *Security*. Retrieved from Guide to the General Data Protection Regulation (GDPR): https://ico.org.uk/for-organisations/guide-to-the-general-data-protection-regulation-gdpr/security/

Ki-Aries, Duncan; Faily, S. (2017). Persona-centred information security awareness. *Computers & Security, 70* (9), 663–674.

Ohlden, A. (2010, 10). Landmark resolution passed to preserve the future of privacy. *Science20*.
Oltermann, P. (2017, 2 17). German parents told to destroy doll that can spy on children. *The Guardian*.
ORACLE. (2014). *Security in Oracle ADF: Addressing the OWASP Top 10 Security Vulnerabilities*. Microsoft Inc., Redmond.
OWASP. (2012). Membership Benefits. 1–2. Bel Air: OWASP.org.
OWASP. (2018). *About The Open Web Application Security Project*. Retrieved from OWASP.org: https://www.owasp.org/index.php/About_The_Open_Web_Application_Security_Project#Code_of_Ethics
OWASP. (2018). *Defense in depth*. Retrieved from OWASP.org: https://www.owasp.org/index.php/Defense_in_depth
Ravasi, D.; Schultz, M. (2006). Responding to organizational identity threats: Exploring the role of organizational culture. *Academy of Management Journal., 49* (3), 433–458.
Samonas, Spyridon; Coss, D. (2014). The CIA Strikes Back: Redefining Confidentiality, Integrity and Availability in Security. *Journal of Information System Security, Volume 10*(3), 21–45.
Schneier, B. (1999). Security in the Real World: How to Evaluate Security. *NetSec Conference*. St. Louis, MO: CSI.
Smyth, A. (1907). *The Writings of Benjamin Franklin (1789–1790)*. (Vol. 10). New York: MacMillian.
The Information Commissioner's Office (ICO). (2018). *Guide to the General Data Protection Regulation (GDPR)*. Retrieved from Website: https://ico.org.uk/for-organisations/guide-to-the-general-data-protection-regulation-gdpr/
The Organisation for Economic Co-operation and Development (OECD). (2005). *Privacy*. Retrieved from Glossary of Statistical Terms: https://stats.oecd.org/glossary/detail.asp?ID=6959
Valentine, B. (2017, 5). *Security Is an Organizational Behavior Problem*. Retrieved from www.securityintelligence.com: https://securityintelligence.com/security-is-an-organizational-behavior-problem/
Wright, Ryan; Chakraborty, Suranjan; Basoglu, A., & Marett, K. (2010). Where Did They Go Right? Understanding the Deception in Phishing Communications. *Group Decision and Negotiation, 19*(4), 391–416.

Appendix

Appendix 5a: OWASP Code of Ethics

- Perform all professional activities and duties in accordance with all applicable laws and the highest ethical principles;
- Promote the implementation of and promote compliance with standards, procedures, controls for application security;
- Maintain appropriate confidentiality of proprietary or otherwise sensitive information encountered in the course of professional activities;
- Discharge professional responsibilities with diligence and honesty;
- To communicate openly and honestly;
- Refrain from any activities which might constitute a conflict of interest or otherwise damage the reputation of employers, the information security profession, or the Association;
- To maintain and affirm our objectivity and independence;
- To reject inappropriate pressure from industry or others;
- Not intentionally injure or impugn the professional reputation of practice of colleagues, clients, or employers;
- Treat everyone with respect and dignity; and
- To avoid relationships that impair — or may appear to impair — OWASP's objectivity and independence.

Appendix 5b: GDPR Article 9(2), Processing Special Categories

Article 9(2):

(a) the data subject has given explicit consent to the processing of those personal data for one or more specified purposes, except where Union or Member State law provide that the prohibition referred to in paragraph 1 may not be lifted by the data subject;

(b) processing is necessary for the purposes of carrying out the obligations and exercising specific rights of the controller or of the data subject in the field of employment and social security and social protection law in so far as it is authorised by Union or Member State law or a collective agreement pursuant to Member State law providing for appropriate safeguards for the fundamental rights and the interests of the data subject;

(c) processing is necessary to protect the vital interests of the data subject or of another natural person where the data subject is physically or legally incapable of giving consent;
(d) processing is carried out in the course of its legitimate activities with appropriate safeguards by a foundation, association or any other not-for-profit body with a political, philosophical, religious or trade union aim and on condition that the processing relates solely to the members or to former members of the body or to persons who have regular contact with it in connection with its purposes and that the personal data are not disclosed outside that body without the consent of the data subjects;
(e) processing relates to personal data which are manifestly made public by the data subject;
(f) processing is necessary for the establishment, exercise or defence of legal claims or whenever courts are acting in their judicial capacity;
(g) processing is necessary for reasons of substantial public interest, on the basis of Union or Member State law which shall be proportionate to the aim pursued, respect the essence of the right to data protection and provide for suitable and specific measures to safeguard the fundamental rights and the interests of the data subject;
(h) processing is necessary for the purposes of preventive or occupational medicine, for the assessment of the working capacity of the employee, medical diagnosis, the provision of health or social care or treatment or the management of health or social care systems and services on the basis of Union or Member State law or pursuant to contract with a health professional and subject to the conditions and safeguards referred to in paragraph 3;
(i) processing is necessary for reasons of public interest in the area of public health, such as protecting against serious cross-border threats to health or ensuring high standards of quality and safety of health care and of medicinal products or medical devices, on the basis of Union or Member State law which provides for suitable and specific measures to safeguard the rights and freedoms of the data subject, in particular professional secrecy;
(j) processing is necessary for archiving purposes in the public interest, scientific or historical research purposes or statistical purposes in accordance with Article 89(1) based on Union or Member State law which shall be proportionate to the aim pursued, respect the essence of the right to data protection and provide for suitable and specific measures to safeguard the fundamental rights and the interests of the data subject.

Chapter 6: Protection Policies and Privacy Notices

At a Glance:

- ❖ Introduction to protection policy
- ❖ The COBIT 5 Framework
- ❖ Protection Policies
- ❖ Privacy Notices: Qualities and Types

Learning Objectives: *Students should be able to…*

- ❖ Understand the nature and formulation of organisational policy.
- ❖ Understand the fundamental aspects of the COBIT 5 framework.
- ❖ Explain the principles and enablers of the COBIT 5 Framework.
- ❖ Explain the types privacy notices and their application.

Key Terms

1) Policies and Procedures
2) COBIT 5: Enablers and Principles
3) Data Protection Policies
4) Privacy Notices

Introduction

An organisation's policies and procedures (PaP's) are akin to the law, and form part of the governance function. They are the rules and regulations governing and guiding the; why, how, when, where, and who of doing what needs to be done to meet the goals and legal obligations of the organisation. The primary function of these documents is to clearly and effectively communicate the reasons for these rules, and the responsibilities of the relevant members of staff when adhering to them.

Policies are high-level documents which set out the overall purpose, functions, and roles relevant to the area being discussed. Policies arise from a management perspective and must have the support and approval of senior management to be effective. PaP's need to be reviewed regularly, to allow for changes in regulation or organisational goals.

Procedures set *how* those rules are to be executed within the organisation to implement the policy. Procedures are written by the team responsible for the specific area relevant to the policy. Procedures must also be approved by senior management or the policy owner.

It is extremely important that these policies and procedures are conceived and written as separate documents, as any changes to one or the other usually requires review and approval by senior management. Where policies and procedures are combined into one document, any changes to one part means the whole document must be scrutinised before approval. This is time consuming and leads to delays in implementation, for what may be a minor procedural change.

When creating PaP's it is important to keep in mind the intended audience. This is an exercise in communication and the goal is to enable staff to comply with the policies and procedures. User-friendly documents will have at least the following:

- *Consistent structure:* which allows staff to quickly and easily locate those PaP's that are most relevant to them. Access controls should be used to limit documents to relevant roles only.
- *Statement:* helping staff to *understand* why the policy exists and how it benefits them.
- *Responsibilities:* details of where and with whom they lie, and what is required of staff.

- *Scope:* to whom it applies; or when it must be applied.
- *Definitions:* terms must be explained clearly, including what is and is not meant, to minimise misinterpretation or confusion. Examples that are relevant to the PaP are useful.
- *Attachments:* including any other relevant documents as links or attachments makes it easier and more efficient for staff to comply.
- *Title:* clear and unambiguous titles to help staff correctly access relevant PaP.
- *Key dates:* showing date of approval or adoption, and the dates of any subsequent review, helps staff to stay current and increases confidence in the PaP's relevance to them.
- *Policy code:* version control must be exercised for each review or change to the document.

Policy Framework: COBIT 5

The Information Systems Audit and Control Association is known as ISACA and is the internationally recognised professional body which deals with IT governance. The COBIT 5 framework was developed by ISACA and is widely used in the practice of good governance in enterprise IT. Developed for use in organisations of all sizes, COBIT 5 is used to align IT with the goals of the organisation and to comply with regulation. There is a strong emphasis on the fundamental role policies play in the implementation of good governance, which enhances the effectiveness of the organisation on all levels. As a single integrated framework, it aligns well with other relevant governance frameworks, and offers complete end-to-end coverage in organisational governance.

Considering that Information Technology is an integral part of the functioning of all organisations, regardless of the sphere of activity. The use of a comprehensive policy framework which considers organisational objectives as foundational in creating policy.

COBIT 5, Five Principles

Principle 1: *Meeting stakeholder needs*

Organisations exist to deliver value to stakeholders. They do this through the balancing of three high-level stakeholder needs; creating benefits for stakeholders, efficient use of resources, and optimising risk. Stakeholder needs are met using the 'goals cascade', which translates high-level stakeholder needs into 'Enterprise' (organisational) goals and then into specific IT goals, these are then mapped to the 7 'enablers'.

Principle 2: *Covering the enterprise end to end*

All functions and processes within an organisation are dealt with under COBIT 5. Information, and related technology are considered as assets of the organisation and should be treated as such by all stakeholders. Governance and management of IT is considered to be enterprise wide; meaning that it should consider and engage with every part of the organisation and with any external parties that have anything to do with the governance and management of the organisation's IT.

Principle 3: *Applying a single integrated framework*

COBIT 5 is a high-level framework which easily aligns with other more detailed frameworks.

Principle 4: *Enabling a holistic approach*

To enable and support comprehensive governance several interactive elements must be assessed within an organisation. COBIT 5 has 7 Enablers; these help the enterprise achieve its objectives.

Principle 5: *Separating governance from management*

Governance is not management. Governance is focused on the evaluation of stakeholder needs, their conditions, and options, in order to set organisational objectives which, deliver stakeholder value. It sets the direction and monitors compliance and performance to the agreed upon objectives and direction. Management plan the implementation, build systems, operate and monitor the activities which are based on the agreed objectives. Their role is to help achieve the objectives set by governance.

Governance and management engage in different activities and use different organisational structures to achieve their goals.

Figure 6.1: 5 principles of COBIT 5

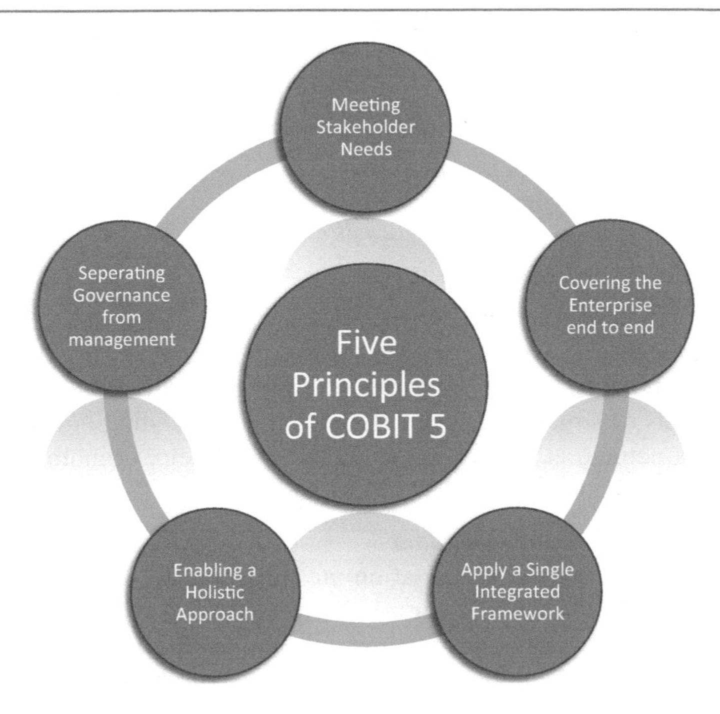

These five principles help organisations to create a holistic framework for governance and management of every part of the organisation and its IT systems. The five principles are supported by seven 'enablers'.

Seven Enablers in COBIT 5

Enablers are the interactive factors which influence the success of a desired outcome. This influence is exercised individually and collectively to help ensure the meeting of goals.

1. *Principles, policies and frameworks:*
These are the tools which help develop the practical guidance for staff to achieve the objectives set by governance, such as the COBIT 5 principles (figure 6.1). This is a day-to-day management tool that outlines what the desired outcomes are and how they can be met to shape the behaviour of the organisation.

2. *Processes*

They are the description of organised groups of activities and practices that achieve specific objectives. Processes should produce outputs which support overall goals in IT related processes within the organisation.
There are 37 COBIT processes grouped under 5 headings:

1) Evaluate, direct, and monitor.
2) Align, plan, and organise.
3) Build, acquire, and implement.
4) Deliver, service, and support.
5) Monitor, evaluate, and assess.

3. *Organisational structures*

Who are the key decision makers within the organisation? These may be individuals or structures, or departments. Understanding how the organisation is structured is vital to determine who will help, or inhibit, the success of the desired outcome.

4. *Culture, ethics and behaviour*

The people within the organisation are important factors in the successful attainment of goals. Often their role is under-rated, and the impact of the culture, ethics, and behaviour of the organisation is underestimated. Therefore, a focus on what the correct ethics and behaviour should be and how these develop into the culture of the organisation ensures progress toward achieving organisational principles and goals.

5. *Information*

For the good governance and smooth running of the organisation, good quality information is required. From an operational perspective, information may be the product or asset.

6. *Services, infrastructure and applications*

Hardware and software technology which provides the services and processing capacities of the organisation.

7. *People, skills and competencies*

Human Resources, the people and the skills and abilities they bring to the organisation. Success in the desired goals and outcomes cannot be attained without understanding what the individuals within the organisation can, and cannot, achieve to meet those goals.

Together, the principles and enablers allow an organisation to align its IT investments with its objectives to realise the value of those investments (represented in figure 6.2).

Figure 6.2: Enablers of COBIT 5

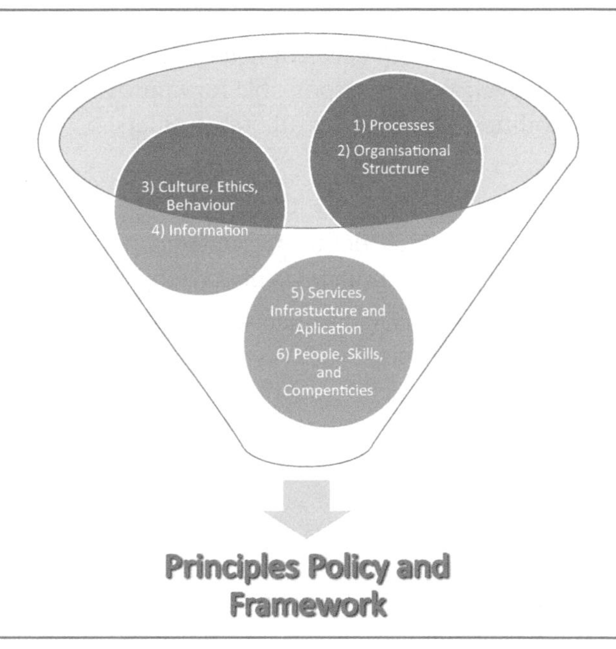

COBIT 5: Principles, Policies and Frameworks in Depth

There are four dimensions to this COBIT enabler;

Stakeholders: these may be defined as policy *creators,* and policy *users.* Policy creators define and set the principles of the psolicy and are usually part of the senior management. Policy is defined and set by considering and analysing the following;

- Governance principles within the organisation.
- Organisational culture.
- Business direction.
- Internal factors.
- External factors.

Once these areas have been analysed the creation of a policy framework must include the core components which meet the policy user's needs. These core components are:

Definitions:

- Scope of the policy;
- Stakeholders it applies to;

- Method for monitoring compliance;
- How to deal with exceptions.
- Determination: Consequences of non-compliance.
- Policy approval: Who is responsible for approving the policy.
- Responsibilities: Outlined for all Stakeholders.

Goals:

These are statements which define the desired outcome from the policy. Goals should arise from the principles underlying the policy and enhance:

- Day-to-day operations and projects;
- Regulatory compliance;
- Business Continuity Management;
- Understanding of scope and authority of roles;
- Protection of intellectual property.

Life-cycle:

The policy life-cycle has six phases which use the information gathered in the previous stages, Stakeholders and Goals, to create, implement, and analyse a dynamic framework.

Phase 1: *Plan*

The objective of the plan phase is to clarify and improve the implementation of the policy principles. The structure of the documentation is decided; one that supports a logical and clear understanding of the policy being communicated. A GAP analysis may be conducted where policies are already in place to help assess and improve the current framework.

Phase 2: *Design*

The design phase has two activities which occur before any writing takes place. The first activity is the priorities setup; this involves the identification of the actual policy to be written, the setting of priorities for the creation of the policy, and the deadlines or review schedule.

The second activity prior to writing is the defining of the structure for the writing, review, and approval of the policy. Essentially, this is approached in five parts:

1) Draft: who will be responsible for researching and writing the policy?
2) Review: an independent review of the draft will improve the quality and the credibility of the policy. Identify who will be responsible for conducting the review.

3) Style: consistent writing style and formats improve clarity and ease of communication. Define the writing standards to be used in-line with the organisation's general policy standards.
4) Approval: determine the procedure for final approval from the person identified in the Stakeholders dimension.
5) Communication and distribution: define the strategy for sharing the policy and training.

Phase 3: *Implementation*

The transition into compliance for a new policy is gradual and requires activities that will encourage and reinforce accountability for the new policy. Enforcement is necessary during the implementation phase, as new habits must be learned.

Phase 4: *Operate*

After implementation, compliance to the policy should become part of the organisation's day-to-day operations. At this stage the policy requirements become part of the organisational culture and contributes to meeting the organisation's goals.

Phase 5: *Monitor/Evaluate*

Monitoring the effectiveness of the policy implementation allows for an evaluation of the degree to which it supports the organisation's goals. The results are shared with the stakeholders.

Phase 6: *Update/dispose*

Policies are reviewed regularly to keep them relevant and effective. Where organisational goals or business direction change, policies must be updated. When policies are found to be obsolete or unworkable then they are disposed of.

Good practices:

Within the organisation management should be separated from governance. This gives rise to the need for additional documentation in support of the policy implementation and management. These supporting documents are often in the form of:

Standards: these are related to the specific policy and designed to enhance the effectiveness of that policy. Standards are often industry-wide, or platform specific, and must be conformed to.

Procedures: the step-by-step actions that must be followed to comply with the policy.

Guidelines: offer additional recommendations which are not mandatory but are considered useful or helpful.

The Data Protection Policy

Article 24 Responsibility of the controller:

1. *Taking into account the nature, scope, context and purposes of processing as well as the risks of varying likelihood and severity for the rights and freedoms of natural persons, the controller shall implement appropriate technical and organisational measures to ensure and to be able to demonstrate that processing is performed in accordance with this Regulation. Those measures shall be reviewed and updated where necessary.*
2. *Where proportionate in relation to processing activities, the measures referred to in paragraph 1 shall include the implemetion of appropriate data protection policies by the controller.*

Article 24

The GDPR states that one of the appropriate organisational measures is the implementation of data protection policies. Having considered the purpose of policies and procedures and the nature of a policy framework, let us look at the drafting process itself. This task is undertaken by the policy owner or writer whose main goal is to create a document that is clear and easily understood by the readers. The policy seeks to align the principles and goals of data protection with the organisation's operational outcomes. It does this through the setting of expected behavioural standards and in communicating the responsibilities, and roles, of those to whom it relates.

Writing concisely forces the writer to distil the information to its most important elements. The temptation to write more than is needed must be overcome, so that only the specific point is made. This avoids confusion or misinterpretation by non-specialists in the field and highlights the need for descriptions or definitions of terms. Remember, clarity of communication is paramount! In practice this applies to words such as "could" and "should";

or any words that indicate flexibility in the choice of action. Words must indicate what is intended and vagueness of any sort is to be avoided.

A simple test is to ask: What would an educated person on the street understand from the document? If the answer is; "not much", then reword the document!

If additional statements are used to clarify the original point, then be aware of the possibility that the original statement has been altered or diluted. For example, a statement such as:

"Staff must not use external devices at their workstations.", implies that away from their workstation, they can (and will) use external devices. Or alternatively, that individuals who are not staff *are* permitted to use external devices.

Policy Document Structure

The structure of the policy will be informed by the existing framework used within the organisation. However, it is useful to have a comprehensive, generic, structure with which to compare and enhance the current model. Below is a brief outline of the necessary components for a data protection policy document.

Title:

In as few words as possible state the specific purpose of the document.

Classification Label and Policy Number

As discussed previously[1], the classification of information is an important part of the data protection process, and policies must be given labels relating to security and organisational area. The policy number and page numbering assists in document version control. Individual requirements should also be numbered as it helps users to find the relevant requirement.

Brief Description

A summary of the policy which does not include any specific details, roughly one paragraph at most.

[1] In chapter 5.

Effective From

The date policy came into force or date of review.

Approval

The name and position of the person who approved the policy.

Policy Owner/Contact

The contact details of the policy owner or the person with responsibility for the policy.

Supersedes

If the policy replaces other policies, name them here, if no policies are superseded then mark as not applicable.

Last Reviewed or Updated

Include any review or revision dates.

Version Control

Any changes and amendments should be documented to improve transparency and clarity. Changes must be communicated to users.

Scope

Specify the scope to whom the policy applies. Name the roles or individuals or departments who must comply with the policy.

Purpose of the Policy

Explain the reasons for the creation of the policy, from a stakeholder benefit viewpoint. The reasons may be based in legal requirements or management goals, but the emphasis should be on the problem-solving nature of the policy. How will it help the organisation to better achieve its goals? Historical background is not needed here.

Risk Appetite Statement

Define the consequences and impact of non-compliance with the policy. Accountability must be highlighted for users to understand the importance to them and the organisation of the policy.

Introduction

Describe the process undertaken for the development of the policy. Explain who was involved and how the final provisions were reached.

Policy Statement

The body of the policy gives the specific principles, goals, and requirements for the primary audience, those who must comply. It must define:

- Who is the policy for, the primary audience?
- What are the requirements, the rules which must be followed? These will be numbered with separate headings to facilitate ease of use. These are *not* procedural 'how to', rather rules of 'what to'.
- Are there restrictions or exceptions to the policy?
- When does the policy not apply?
- What are the roles and responsibilities relevant to the policy?

Definitions

List terms and acronyms that are unfamiliar to non-specialists.

Related Policies, Procedures, Forms, Guidelines, and Other Resources

Policy works in interconnection with other policies, and related policies and other documents should be included or linked to. These may be internal links or supportive links from external sources; such as the Supervisory Authority.

Data Protection Privacy Notice

The GDPR stipulates that individuals must be provided with easily accessible information on the collection, storage, sharing, and destruction of their personal data. The privacy notice is the most widely used format for presentation of this information to individuals. Unlike a privacy *policy,* which is a document internal to the organisation, a privacy *notice or statement,* is external to the organisation in that it is a public document containing statutory information. Failure to provide this document, which contains the statutory privacy information, carries the risk of court action by the data subject against the organisation, or enforcement action on the part of the Supervisory Authority.

Privacy notices must be written in clear, concise, and easy to understand language which is in no way misleading, dishonest or ambiguous. The GDPR states that privacy notices must contain the following:

1) *Purpose for processing:* why is the data being collected? All reasons for the processing of personal data must be stated; where they are not so stated the undisclosed purposes may be considered unlawful.
2) *Lawful basis for processing:* one of the six lawful bases[2] must be stated at least.
3) *Legitimate interest:* the data controller and third-parties must state what legitimate interest is being applied in order to further process data beyond the purpose and lawful basis. The data subject's reasonable understanding of what will be done with their personal data informs the correct application of this category within the privacy notice.
4) *Categories:* personal and sensitive data categories, which are collected or processed etc, must be explained in a way that the data subject can easily understand.
5) *Recipients of personal data:* who will receive the data and why is it shared with them? Which third-party groups, organisations, and categories of recipients receive the data must be made explicit, preferably in list form. Where the reason is to comply with the law, this must be stated, any other reason, commercial or otherwise, must be made clear.
6) *Third country transfers:* where data is transferred outside the European Economic Area (EEA), the data subject must be told the reasons for the transfer, and how the data will be safeguarded.
7) *Collection:* where and how the data is collected, whether directly volunteered by the data subject, or from a publicly available source, such as the electoral register, or through third-parties. If the provision of the data impacts the fulfilment of a contract, such as a loan application, then the data subject must be made aware of the impact and consequences of withholding their personal data.
8) *Storage and retention:* data must be retained for the minimum period required, and data subjects must be informed as to how that period is determined, and how long it will be.

[2] See Chapter 1: Lawful bases for processing.

9) *Rights of the data subject:* must be highlighted, explained, and include direction for the exercising of those Rights. The right to withdraw consent at will (where consent is the Lawful basis for processing), and the right to complain to the Supervisory Authority are two rights which must be made clear.

Types of Privacy Notices

The ICO has favoured a layered approach in the creation of its privacy notice[3] as this format provides information in 'layers' which are easier for the user to comprehend. The first layer presents the highlights contained in the privacy policy; it presents information in a short and easy to take in style. The points listed in the highlights layer should link to the relevant section of the full privacy policy, this allows the reader to access the full details of the policy. A third, Frequently Asked Questions layer may be provided from either, or both, previous layers. The two layers can be made of from a just-in-time privacy notice and a summary privacy notice.

Summary privacy notices

A summary of the full privacy notice, this format provides some detail under the main bullet point headings. It will usually consist of the headings or sub-headings contained in the full policy document, with a few sentences giving more detail on the heading subject area. Based on the heading suggestions given above, a summary privacy notice would contain the following headings:

- What personal data is being collected?
- Why is it being collected?
- Who is collecting it?
- How is it being collected?
- What will be done with the data?

---

[3]https://ico.org.uk/global/privacy-notice/

- Will it be shared, why and with whom?
- What will third-parties do with the data?
- How will this affect the data subject?
- Will this give rise to reasons for complaint?

Just-in-time privacy notices

The ICO has also endorsed the use of just-in-time privacy notices as a method of providing people with privacy information at the point of collection, throughout a website. These notices appear at the data input point on the data subject's screen. These notices can be in the form of written messages, videos, or other media; and language must remain clear and simple to understand. A consent mechanism, such as a check box, or button may be included in a just-in-time privacy notice. This must explain the right to withdraw consent at any time and include the same functional ease for the removal of consent as that available for giving that consent.

Chapter Review

Summary of Key Points

1) Overview of an organisation's PaP's.
2) The COBIT 5 Framework.
3) COBIT 5's Principles and Enablers.
4) Protection Policy, Article 24, and a template of a policy document.
5) Intro to privacy notices and types of privacy notices.

Cognition

- How should an organisation's Codes of Conduct operate within a policy framework?
- Think about the COBIT 5 framework and how it might relate to different industries.
- What qualities make a good privacy notice, which type would you use?

References

Amason, A. (2017). Distinguishing the Effects of Functional and Dysfunctional Conflict on Strategic Decision Making: Resolving a Paradox for Top Management Teams. *Academy of Management Journal, 39*(1).

Carrillo, J. (2013). IT Policy Framework Based on COBIT 5. *ISACA,* 1.

De Haes, S., Van Grembergen, W., & Debreceny, R. (2013). COBIT 5 and Enterprise Governance of Information Technology: Building Blocks and Research Opportunities. *Journal of Information Systems, 27*(1), 307–324.

Department for Education. (2013). Data protection: privacy notice model documents. London: UK.gov.

Department for Education. (2018). Privacy notices: An explanation of privacy notices. London: Assets Publishing Service Uk Gov.

Hong, K.-S., Chi, Y.-P., Chao, L., & Tang, J.-H. (2003). An integrated system theory of information security management. *Information Management & Computer Security, 11*(5), 243–248.

iapp. (2016). ICO endorses use of 'just-in-time' notices R. *Europe Data Protection Digest*(10).

ICO. (2016). CONTROLLER AND PROCESSOR SECTION 1 GENERAL OBLIGATIONS. *GDPR recitals and articles.*, 54–55. ICO website.

ICO. (2018). *Controller's contact details.* Retrieved from Privacy notice: https://ico.org.uk/global/privacy-notice/controller-s-contact-details/

Ko, D., & Fink, D. (2010, 10 19). Information technology governance: an evaluation of the theory practice gap. *Corporate Governance: The international journal of business in society, 10*(5), 662–674.

Kulkarni, G. (2017, 4). Applying the Goals Cascade to the COBIT 5 Principle Meeting Stakeholder Needs. *COBIT Focus.*

Lainhart, J., & Oliver, D. (2010). Integrating ISACA Frameworks Into One Overarching Framework: COBIT 5. *COBIT Focus, 2010*(2), 6–9.

Menevse, A. (2011). Toward Better IT Governance With COBIT 5. *COBIT Focus, 4*, 1–4.

Munur, Mehmet; Branam, Sarah; Mrkobrad, M. (2012, 9). Best Practices in Drafting Plain-Language and Layered Privacy Policies. *iapp News.*

Office of Policy and Efficiency. (2018). *User Guide to Writing Policies.* University of Colorado, Boulder.

Office of Policy and Efficiency. (n.d.). *User Guide to Writing Policies.*

Weill, P. (2004). Don't just lead, govern: How top-performing firms govern IT. *MIS Quarterly Executive, 8*(1), 1–21.

Wolden, Mark; Valverde, Raul; Talla, M. (2015). The effectiveness of COBIT 5 Information Security Framework for reducing Cyber Attacks on Supply Chain Management System. *IFAC-PapersOnLine, 48*(3), 1846–1852.

Chapter 7: DPO, DPIA, and DSAR

At a Glance:

- Appointing a DPO
- Tasks of the DPO
- Characteristics of a good DPO
- DPIA: Article 35, legal requirements, and conducting a DPIA
- DSAR: The data subject and the role of the organisation

Learning Objectives: *Students should be able to…*

- Explain the criteria for appointing a DPO.
- Understand the qualities of a good DPO.
- Identify the legal requirements of a DPIA and the how it builds into the organisations risk policy.
- Describe the process of conducting a DPIA and the questions an organisation should be asking.
- Explain the data subject's right of subject access.
- Recognize how a data subject can request access to their personal data and the responsibility on the organisation.

Key Terms

1) Data Protection Officer
2) Data Protection Impact Assessment
3) Data Subject Access Request
4) Prior Consultation

Introduction

This chapter is intended to give a better understanding of the key elements that must be implemented before an organisation can fully comply with the GDPR. The chapter will cover the 3 D's of GDPR compliance: The Data Protection officer (DPO), a Data Privacy Impact Assessment (DPIA), and a Data Subject Access Request (DSAR). These last two key steps interrelate with the obligations and role of the DPO, it is important to realise that not all organisations require a DPO. However, all organisations should be able to complete a DPIA and must be able to locate where personal data is held, should a DSAR be filed by one of their customers or data subjects. The data that an organisation holds, and how it should be managed, is covered in chapter 3.

There are many steps required to comply with the regulation, and these 3 steps act as an anchoring point for any practitioners of data protection and for organisations seeking to find best practice models of operation when it comes to adequate data protection measures. The previous two chapters are fundamental in understanding the 3 D's of compliance. Without a thorough understanding of data protection by design and default (DPDD) these elements, the 3 D's of the GDPR, might not make much sense for departments within an organisation. It is important to build security into the foundation of the organisation, making it a state of being rather than a tick-box exercise; the 3 D's, policies, and privacy notices are fundamental to this.

The next few sections of this chapter will outline the key role of a DPO, and the circumstances under which it is appropriate to appoint one, referencing the legislation. The purpose of the DPIA, and when, and how often, an organisation should conduct a DPIA. Finally, a description of a DSAR and its function, the circumstances under which a data subject can request a DSAR, and how an organisation should respond to a DSAR.

Data Protection Officer

The DPO is an independent entity that ensures the organisation is adhering to the GDPR. The primary goal of the DPO is to protect the personal data of EU citizens that the organisation is processing. Under the GDPR, the DPO is cited in Article 37, Article 38, and Article 39 parts of these articles will be cited in the relevant sections the articles mentioned outline the designation of the DPO, the position of the DPO, and the tasks of the DPO. In brief, the DPO should work to ensure that the organisation is complying within the framework of the regulation, whilst also not inhibiting the operations of the organisation. In this sense the DPO should not be a yes-man or a naysayer but rather work with the organisation to find a solution to the problem when presented with one.

The DPO should be offer advice for data protection, they do require a varied skilled set, with technical knowledge of the law to fully understand the regulation. Above all the most important element for a DPO is to act as an individual, there should be no conflict of interest with the organisation and the DPO should be confident enough to undertake the tasks afforded to them when a situation surrounding data protection arises.

Appointing a DPO

Appointing a DPO is a pivotal role in an organisation, it can mean the difference in a heavy penalty or a simple warning. The DPO must be flexible, and act as an arbitrator between the law and the actions of the organisation. A good DPO acts as a guardian of the spirit of the law, much like the goddess Athena who brought balance to the Greeks with the law, they act with the interest of the people of Europe by balancing their data rights without disrupting the balance within the business world or society at large. Therefore, the process of appointing a DPO carries a burden of responsibility to do it right. The first question an organisation should be asking is weather or not they require a DPO.

Although the description above might suggest that a good DPO would be well placed in any organisation, the regulation does not require an organisation to appoint a DPO unless one of the criteria is fulfilled within Article 37(1):

The Controller and the Processor shall designate a data protection officer in any case where:

1) *the processing is carried out by a public authority or body, except for courts acting in their judicial capacity*
2) *the core activities of the Controller or the Processor consist of processing operations which, by virtue of their nature, their scope and/or their purpose, require regular and systematic monitoring of Data Subjects on a large scale; or*
3) *the core activities of the Controller or the Processor consist of processing on a large scale of special categories of data pursuant to Article 9 and personal data relating to criminal convictions and offenses referred to in Article 10.*

Article 37(1)

The regulation outlines the scenarios in which a DPO is required. The first point refers to a public authority, which in chapter 1, was covered as a lawful basis for processing. The primary reference to the second point is to "systematic monitoring of Data Subjects (DS) on a large scale", this refers to the process within the organisation. The systematic monitoring of data subjects personal data on a large scale, requires the organisation to appoint a DPO regardless of the size of the organisation, there are no exemptions for SME's.

An example of this is if the organisation's HR department holds large amounts of personal data on the organisation's employees across Europe. This would fulfil the second criteria of the article, and as a result would require a DPO, to facilitate compliance with the regulation. This is also the case for organisations dealing with high volumes of customer data, where the systematic monitoring is required for the ongoing services they provide. Lastly, the same obligation to appoint a DPO is required if the organisation processes many special categories of data (flow chart for criteria of appointment found in figure 7.1). Special categories of data are covered in Article 9 of the GDPR: a few descriptors of special categories of data are shown below:

- Racial or Ethnic Origins.
- Religious or Philosophical Belief.
- Trade Union Membership.
- Biometric Data or any Data Concerning Health.
- Natural Person's Sex Life or Sexual Orientation.

These are a few of the special categories of data outlined in the regulation. The regulation also prohibits any processing of the special categories of data unless certain conditions are met, which are named in Article 9(2), this is detailed in chapter 5. If the conditions are met and the organisation can lawfully process the special categories of data, they will be required to appoint a DPO, given their core activities require processing a high volume of that data.

Appointing a DPO is a crucial step towards compliance. The DPO plays the role of the navigator of the organisation's ship through the sea of regulation. The DPO should assist in fostering good business practice, whilst upholding the rights of EU citizens, and balancing the needs of the business with those of the people of Europe. Therefore, the DPO must remain neutral, and should not face sanctions for upholding the principles of the GDPR. The nature of the positions requires a DPO to exhibit certain characteristics to fulfil the responsibilities of the role.

Figure 7.1: DPO Appointment Criteria Flow Chart

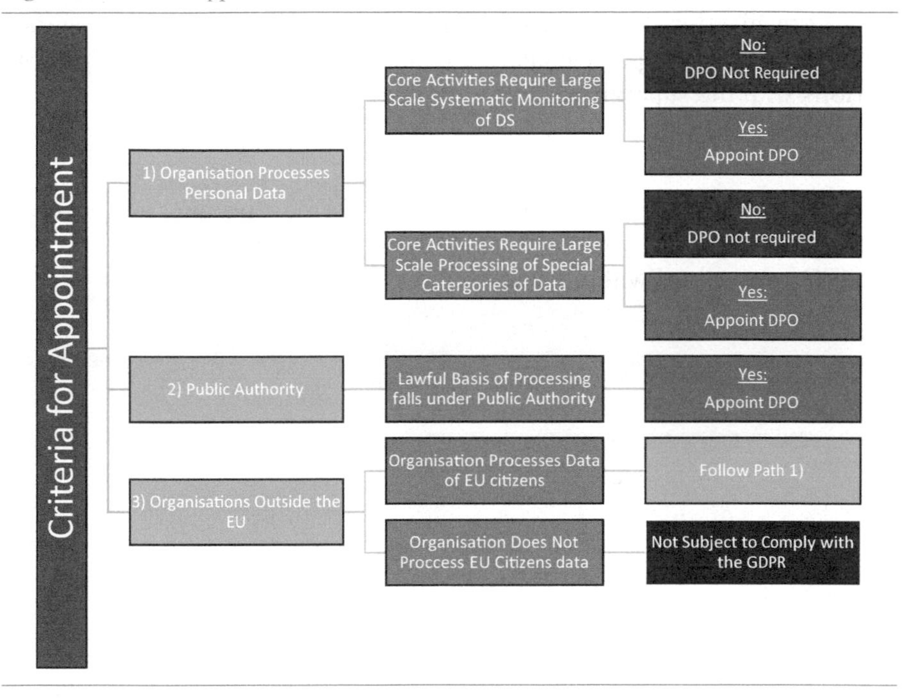

What Makes a good DPO?

As mentioned above, the most important element of the DPO is to remain independent and aligned solely to the requirements of the law they uphold, whilst balancing the rights of EU citizens and the needs of the organisation. This requires the DPO to show a great sense of responsibility; a responsibility that comes from, understanding the potential consequences that, access to high volumes of personal data could create. The DPO should uphold the values of confidentiality, integrity and ethics within themselves to accomplish what others are not willing to do.

In the event that the organisation significantly falls short of the requirements for compliance, the DPO has a duty to report this failing, to the supervisory authority, in a timely manner. Therefore, the DPO must show a degree of resilience, as organisations may take some time to adjust to change in the daily operating of the business. The DPO may face resistance by top-level management, whose primary concern is bottom-line impacts to the organisation. At the same time a good DPO should be able to teach the organisation the benefits arising from compliance with the regulation, to facilitate a perspective change within management fostering good relations with the supervisory authority and the people of Europe.

Tasks of the DPO

Now that the criteria for appointment and the positive qualities of the DPO have been covered, you may be wondering what a DPO does? The task of the DPO are outlined within the regulation

1) The data protection officer shall have at least the following tasks:

(a) To inform and advise the Controller and Processor and the employees who carry out processing of their obligations pursuant to this Regulation and to other Union or Member States data protection provisions;

(b) To monitor compliance with this Regulation, with other Union or Member State data protection provisions and with the policies of the Controller or Processor in relation to the protection of personal data, including the assignment of responsibilities, awareness-raising and training of staff involved in processing operations; and the related audits;

(c) To provide advice where requested as regards the data protection impact assessment and monitor its performance pursuant to Article 35;

(d) To cooperate with the Supervisory Authority:

(e) To act as the contact point for the Supervisory Authority on issues relating to processing, including the prior consultation referred to in Article 36, and to consult, where appropriate, with regard to any other matter.

Article 39(1)

Article 39(1) sets out the primary tasks of the DPO.

1) (a) Refers to the DPO as an educator; the DPO must inform the organisation of the rules set in place by the Regulation. This would only apply where the employees, Controller, or Processor carry out the processing of persona data.

1) (b) The DPO has an obligation to monitor the organisation's behaviour and analyse the alignment of internal policy with the compliance requirements of the Regulation. It is also required that policy must adhere to Union or Member State data protection provisions; meaning not only the Member State the organisation operates in, but within any other Member States the organisation processes the personal data of any EU citizens.

1) (c) Directly related to the DPIA, discussed later in the chapter, the DPO must advise top-management on the areas of danger within the data network and the level of protection of the Data Subjects.

1) (d) The DPO must cooperate with the Supervisory Authority. For example, in the UK, if the organisation has suffered a data breach, the DPO must inform the ICO in a timely manner.

1) (e) The DPO must have a relationship with the Supervisory Authority, and assist it in any way necessary, relating to the processing of data and data protection. Article 36 refers to the need for prior consultation in situations with high risk if the Controller has not taken steps to mitigate the risk.

The tasks of the DPO limit who should be appointed DPO, an example of this: an organisation should not appoint one of the department heads (HR etc.) as a DPO, especially if the department head deals in the processing of personal data as this creates a conflict of interest and would not comply with the Regulation.

Among the tasks that the DPO must take responsibility for, the DPO must also hold a position equal to that of top-level management within the organisation. There must be free and fast dialogue between, the DPO and the Supervisory Authority and, the DPO and top-level management of the organisation.

For further clarification the DPO is not the same as the organisation's Chief Information Officer (CIO), the Chief Information Security Officer (CISO), or the Chief Privacy Officer (CPO). Appointing any of the aforementioned personnel would result in a conflict of interest. The reason for this is that the information team has a direct influence on the purpose and lawful basis for processing, and it is they that choose how the organisation uses the data in a business sense rather than a legal sense.

A more appropriate independent entity that could fulfil the role of the DPO, would be the organisation's; Chief Compliance Officer, a member of the audit team, external contractors such as consultants, or any other reporting entity that can remain independent of the organisation. The organisation should work towards formulating internal policy, that can facilitate the independence of the DPO that they will appoint. The internal rules should include safeguards within the policy to ensure the DPO's position and function within the organisation, along with any other job description, to avoid any conflicts of interests.

In summary the appointment of a DPO, is not the smoking gun which proves compliance. Ultimately it is the organisation's responsibility to comply with the Regulation, not the DPO's. it is Clarification of the organisation's obligation to support the DPO was provided in the Recitals[1] to the GDPR, in ways that have been mentioned before, but in brief:

- The organisation's senior management must support the DPO (Board level/C-level).
- The organisation must give sufficient time to the DPO to fulfil their duties.
- The organisation must support the DPO through financial, infrastructure, and staff resources.
- The organisation must publicly communicate the position to all employees.
- The DPO must have access to all data processing activities (HR, Legal, Marketing, etc.)
- DPO must be in continuous training with the organisation and finally;
- The DPO may require a team, depending on the size and structure of the organisation.

Data Protection Impact Assessment

The data protection impact assessment (DPIA) is cited in Article 35 of the GDPR. A data protection impact assessment assists the organisation by identifying risks associated with new business operations, processes, products, or services. This process concerns the risk to the rights and freedoms of the natural persons and acts in the same way as a traditional risk assessment would. The DPIA of the GDPR is not a new concept, organisations across the globe,

[1] Article 29 Working Party provided *Recitals* to clarify the GDPR, where needed.

have been using similar methods of risk assessment prior to DPIA's, which were known as a Privacy Impact Assessment (PIA). When an organisation wishes to undertake a new project, adopt a new system, or introduce a new technology, a DPIA will have to be conducted, which analyses the privacy risk, to the natural persons, that the project or system may exhibit if implemented. A DPIA can demonstrate an organisation's progress towards protection by design (discussed in chapter 5), and it is a great tool to assist the DPO in fulfilling their duty, but most importantly it is required by law in certain circumstances.

There is a legal obligation to conduct a DPIA for new projects and systems, and there are benefits to conducting a DPIA. It can help the organisation make better decisions and improve management of data protection risks. A DPIA can improve how an organisation uses information: it facilitates the 1st and 2nd principles of the GDPR, fair and lawful processing, and purpose limitation (explained in chapter 1)[2].

Furthermore, a DPIA improves transparency where the data subjects can understand how and why their personal data is being used, which can give an organisation a competitive advantage by building trust with their customer base. A DPIA can answer organisational questions such as how is the data going to be used? Will it be cost effective? What are the risks associated with the project? To name a few.

There is also a financial benefit of a DPIA, also fulfilling the 3rd principle[3] of the GDPR; data minimisation, through minimising the data the organisation collects. Any problems that might develop later in the project, can be more easily dealt with, potentially requiring less costly solutions. The fundamental framework of a DPIA gives both the organisation, and the data subject, assurance that the new endeavour is managed with best practice principles.

Legal Requirements

The legal requirements for a DPIA are outlined in Article 35 of the GDPR. In general, a DPIA should be carried out when the new project, particularly ones that involve new technologies, may involve high risk to the rights and freedoms of Natural Persons. The Controller must conduct a DPIA prior to

---

[2] *Reminder:* Principle #1: data must be processes lawfully, fairly, and transparently. Principle #2: the data must be processed in so far that it achieves its legitimate purpose. Conducting a DPIA will help an organisation accomplish the first two principles set out in the GDPR.

[3] *Reminder:* Principle #3: Minimisation means collecting only that data which can and will be used for the purpose specified.

the processing of that data. There are more specific cases in which a DPIA must be conducted which are mentioned in Article 35(3):

3) A data protection impact assessment referred to in paragraph 1 shall in particular be required in the case of:

(a) a systematic and extensive evaluation of personal aspects relating to Natural Persons which is based on automated processing, including profiling, and on which decisions are based that produce legal effects concerning the Natural Person or similarly significantly affect the Natural Person;

(b) processing on a large scale of special categories of data referred to in Article 9(1), or of personal data relating to criminal convictions and offences referred to in Article 10; or

(c) a systematic monitoring of a publicly accessible are on a large scale.

Article 35(3)

The final points of the Article, which were not mentioned above, relate to the Supervisory Authority. The Supervisory Authority shall publish a list, to be made publicly available, about operations that require a DPIA. They may also choose to create a list in which new operations would not require a DPIA, although this is at the discretion of the Supervisory Authority.

In the UK, under Article 35(4)[4], the ICO (UK Supervisory Authority) has published a list of operations that would automatically require a DPIA, below is that list the ICO has created:

1) New Technologies: processing involving the use of new technologies, or the novel application of existing technologies (including AI).

[4] The Supervisory Authority shall establish and make public a list of the kind of processing operations which are subject to the requirement for a data protection impact assessment pursuant to paragraph 1. The Supervisory Authority shall communicate those lists to the Board referred to in Article 68. - Article 35(4).

2) Denial of Services: Decisions about an individual's access to a product, service, opportunity or benefit which is based to any extent on automated decision-making (including profiling) or involves the processing of special category data.
3) Large-scale Profiling: any profiling of individuals on a large scale.
4) Biometrics: any processing of biometric data.
5) Genetic Data: any processing of genetic data, other than that processed by an individual GP or health professional for the provision of health care direct to the data subject.
6) Data Matching: combining, comparing, or matching, personal data obtained from multiple sources.
7) Invisible Processing: processing of personal data that has not been obtained directly from the data subject, in circumstances where the controller considers that compliance with Article 14[5] would prove impossible or involve disproportionate effort.
8) Tracking: processing which involves tracking an individual's geolocation or behaviour, including but not limited to, the online environment.
9) Targeting of Children or Other Vulnerable Individuals: the use of the personal data of children, or other individuals, for marketing purposes, profiling or other automated decision-making, or if you intend to offer online services directly to children.
10) Risk of Physical Harm: where the processing is of such a nature that a personal data breach could jeopardise the (physical) health or safety of individuals.

It should be stated that this list only pertains to the UK Supervisory Authority, the ICO, and that the Supervisory Authority of other Member States are required by the GDPR to publish a list in their jurisdiction. The ICO has yet to publish a list of operations that do not require a DPIA.

Defining Article 35

The GDPR beautifully describes the ways in which organisations achieve good practice but does poorly when defining the meaning of specific actions. A strong case for this comes from the scope of the legislation, the GDPR

[5] Article 14: Information to be provided where personal data have not been obtained from the data subject. This is essentially a check list for the Controller to abide by if data was directly given by the data subject. The Controller must inform the data subject referred in paragraphs 1–6.

attempts to remain as broad as possible to encompass emerging technologies and threats to the rights and freedoms of the Natural Persons. Having said that, there are ways to understand what the GDPR and its articles are trying to convey.

Firstly, in Article 35(1) the term *'systematic and extensive'* occurs frequently throughout the Regulation, but what might that mean for an organisation? The European guidelines on the DPO provisions explains 'systematic' as processing that:

- Occurs according to a system;
- Is pre-arranged, organised or methodical;
- Takes place as part of a general plan for data collection; or
- Is carried out as part of a strategy.

On the other hand, 'extensive' relates to processing that either covers a large area, a wide range of data, or a high volume of individuals.

Similarly, in paragraph 1 of Article 35, the terms *'legal effect'* and *'significantly affect'* are utilised. Again, the GDPR does not do much do define what the effects might be but briefly, a *'legal effect'* is anything that affects a natural person's legal rights or status. A *'significant affect'* is broader and could relate to anything that affects the person's financial, reputational, or health status; or their access to economic or social opportunities.

Finally, the term *'large-scale'*, the GDPR does not outline the prerequisites for an operation to be considered large scale. An organisation might have a better explanation for defining large scale operations, but generally large-scale considers[6]:

- The number of individuals concerned;
- The volume of data;
- The variety of data;
- The duration of the processing; and
- The geographical extent of the processing.

Some examples of this might be; banking companies processing customer data, hospitals processing patient data, or mobile apps processing user geolocation data. This is relevant primarily in the organisational context;

[6] Based on the ICO's interpretation of large scale operations.

individuals who process data, such as a doctor processing data on a single patient, are not considered to be involved in large-scale processing.

Where appropriate, the organisation should consult with the data subject to engage them in the DPIA process. This is not mandatory, but for some organisations it might be necessary to consult data subjects on any changes that might occur to the product or service they provide; especially in cases of high risk to the rights and freedoms of the data subject. This begs the question; what processes should be considered high risk and, when should a DPIA be conducted, based on risk?

Like some of the other terms in the GDPR, risk is not defined. Regarding risk, the organisation should already have a general idea of what risk means to them, but for a DPIA risk specifically refers to risks to the rights and freedoms of natural persons. The most widely used definition of organisational risk comes from the ISO 31000 (covered in chapter 8), which defines risk as the effect of uncertainty on objectives. The first measurement should be; is the event possible and what is the likelihood of it happening relative to the severity of the consequences.

A few examples that would constitute risk under Article 35 and would warrant a DPIA, are privacy risks. Privacy risks should already be registered in the risk register of an organisation, but for further clarification some examples of a privacy risk may be:

- Data is inaccurate or out-of-date
- Data that is kept for too long
- Data lakes (excessive data storage) or irrelevant data
- Data not made secure in the transmission or storage; or
- Data that is disclosed to the wrong people/entity.

These might be considered high risk, and in the event of high risk manifesting due to a new process or operation, a DPIA is not sufficient to implement the new process or operation. If high risk is found to be possible, then prior consultation would be required under Article 36 of the GDPR.

Prior Consultation

Prior consultation, with the Supervisory Authority, is required in the case a DPIA asses high risk to the rights and freedoms of the Natural Person, prior to processing. Where the Supervisory Authority finds that a Data Controller has insufficiently identified or mitigated the risk, the Supervisory Authority will provide written advice to the Controller and where necessary the Processor. The Supervisory Authority may also exercise its power indicated in Article 58. Prior consultation falls under Article 36 of the GDPR. This is related to the Supervisory Authority or the Data Subjects, one example would be if processing affects employees, then it may be necessary to consult with the trade union as well as the Supervisory Authority.

Conducting a DPIA

There is a relatively straight forward process to conducting a DPIA, and this section will examine the steps involved in carrying out a DPIA. The first step is to ascertain if a DPIA is required, the need for a DPIA stems from the risk involved to the rights and freedoms of the Natural Person. The DPIA should show if the risk is present or not, but generally if the operation involves processing high volumes of personal data that is sufficient to warrant a DPIA. Furthermore, identifying the purpose and use for collecting the data, method of collection, and types of data being collected, is part of the pre-planning in identifying the need for a DPIA. A template of a DPIA can be found in the appendix of this chapter.

The second step is to understand the information life-cycle. This is the meat of the pre-planning stage; such as identifying the data points and where the data is coming from, how is it stored, etc. This was covered in chapter 3, data flow mapping, and this tool is essential to conducting a DPIA. Knowing the data and information life-cycle, from collection to destruction of that data, will assist in conducting a DPIA and will also help in identifying the areas of the data flow map that may, or may not, have to be adjusted after the implementation of a new system or process.

Once the data flow and the information cycle have been mapped, the next step is to discover the risks to privacy. This can be achieved by looking at the information life-cycle and identifying gaps or vulnerabilities in the system. For example; if a marketing firm collects data from data subjects using the lawful basis of consent, through telephone marketing, then the collection point is at the call centre. The next question to ask is, where does the data go from there? Is it stored on servers within the call centre? Or is it transmitted to third-party data centres, or cloud storage? Then the organisation must ensure that the risk of those storage services or servers being breached are mitigated. A change in the lawful basis of processing would also be involved in identifying risks.

The next step after discovering the risk(s) to privacy, is to then identify solutions to ensure privacy. The organisation should evaluate which solutions would work best to ensure that the privacy of the data subject is intact. This could mean using encryption on the method of storage, or changes in technical and administrative controls on the data that is being handled, so that only the individuals tasked to process the data can do so.

The last remaining steps involve documenting outcomes and solutions. After identifying the information life-cycle, the risks to the rights and freedoms, and any potential solutions, the organisation's management should sign-off the DPIA and document the process. The organisation should integrate the outcomes of the DPIA into the new project or operation, ensuring that the privacy to the individual is assured. Finally, the organisation should evaluate the results, monitor the level of protection and, send any feedback to the necessary teams and departments.

DPIA quick-step guide

Step 1	•Determine the Need for a DPIA
Step 2	•Understand the Information life-cycle •Have a data flow map ready
Step 3	•Determine the privacy risk •Other risks?
Step 4	•Indentify gaps in the data flow •Find solutions to the privacy risks
Step 5	•Sign-off the DPIA by management or relevant authorites
Step 6	•Integrate the results of the DPIA into the new process or operation
Step 7	•Lessons learned and feedback

Data Subject Access Request

The final section of this chapter will look at the Data Subject Access Request (DSAR). The data subject has a right to know if a data controller is processing their personal data, and if so, the data subject also has the right to access that data, followed by further information surrounding their personal data, such as:

- The purpose of processing.
- The categories of personal data concerned.
- Any third parties that have received your personal data (third countries or international organisations).
- The period in which the data will be stored, or the criteria used to determine that period.
- The right to request from the controller rectification of the data stored (correction such as birthdates, addresses, etc.) or the right to erasure of the data/restriction of processing.
- The right to lodge a complaint with the Supervisory Authority (ICO in the UK)
- If the personal data is not collected from the data subject, they have the right to know from where that data is being collected.
- And finally, the existence of any automated decision-making processes and the logic involved in utilising the automated process.

The DSAR process was previously known as a subject access request (SAR), under the precursor to the GDPR; the EU Data Protection Directive. The GDPR has implemented changes to the criteria in which a data subject can exercise this right; the most obvious change is that organisations can no longer charge a fee for a DSAR, they must provide the data subject with a free copy of the information they hold. However, there are certain circumstances in which an organisation can charge a fee; in cases where the request is 'manifestly unfounded or excessive'. Essentially the data subject cannot bombard the organisation with DSAR's, and it is also within the organisation's right to refuse the request in these circumstances. The organisation can also refuse an access request in the event that the data includes information about another individual, unless the other individual in question has agreed to the request.

How to access the data

The DSAR, is a subject access request which is a right within the GDPR afforded to data subjects if an organisation holds any personal data on the individual, this right is known as the right of access. A data subject can exercise this right by communicating with the organisation either through direct contact or in writing. There are a few steps involved in exercising the data subject's rights as outlined by the ICO:

Step 1:

- *Identify which organisation the request needs to be sent to.*
- *Identify what kind of personal data needs to be accessed (Data types such as addresses, phone numbers, names, dates, messages sent, etc.)*

This step is used to highlight which organisation should be targeted by the DSAR, and the types of data the data subject wishes to access. From the organisation's perspective, it is crucial that a well-defined data flow map to help identify where the personal data is being held and, a system in place in the event a data subject exercises their right to erasure.

Step 2:

- *The data subject should make the request directly to the organisation and need not go through the Supervisory Authority (unless the organisation has failed to process the request in a reasonable time, 40 days stated in the GDPR).*
- *It is at this point that the data subject should identify specifically, the data they are requesting, it should be as clear as possible; to make it easy for the organisation to find exactly what the individual is requesting.*
- *When making a subject access request, the data subject should provide any information that can identify who they are, so the organisation can find the personal data that is being requested, this includes:*
    - *The data subject's name and contact details.*
    - *Information that distinguishes the data subject from any other person with the same name held within the organisations data base (account numbers, usernames/handles, etc.).*
    - *Details or relevant dates that will help find what the data subject requested.*

This step requires a bit of work for the data subject but is important for the organisation, as they can fulfil the request in a timely manner, given all the information provided is correct. The data subject may, for example, request a personnel file that is kept on them, or correspondence from them to a college from a specific date (ex. 7th June 2016 to 7th July 2016). A template of a letter a data subject might send an organisation is provided in the appendix[7].

Step 3:

- *The data subject should keep a copy of the access request.*
- *The data subject should keep any proof of delivery or postage.*

The Organisations Role

If a DSAR is requested as stated in the GDPR, the organisation has a deadline to satisfy the request. The GDPR gives organisations one month to complete a subject access request. As stated previously, the organisation cannot charge a fee to the data subject to cover administrative costs, it must provide the data subject a free copy of the personal data that has been requested.

There are certain situations in which an organisation can refuse a DSAR and these are, situations in which the data subject has requested the information of another individual (this is to protect the other party's privacy), unless the person has given explicit permission to do so. The second is when the request is 'manifestly unfounded or excessive', this can happen if a data subject has repeatedly requested the same information or continues to ask for personal data in many different circumstances.

The organisation is compelled by the regulation to fulfil the DSAR, but is given one month to complete the requests. If the organisation has failed to do so within the one-month time frame, the data subject may request a subject access again, but a complaint should be made to the Supervisory Authority as the organisation has failed to comply with the regulation.

The organisation is within its right to request additional information to determine the identity of the individual requesting the DSAR. The organisation is also responsible to provide additional information, other than the personal data the data subject has requested, this was covered in the first part of this section.

---

[7] ICO template letter for DSAR.

The main challenge of a subject access request is that the GDPR does not state where the request must be directed. This means that the data subject has the right to direct the DSAR to any department within the organisation, nor does the GDPR state a specific way to make a valid request, so an individual can make a subject access request verbally, or in writing (email or post).

This poses an organisational challenge as the DSAR could come to any department. Therefore, it may be in the best interest of the organisation to have a department who is client facing to handle any request that come their way, and the organisation should have some training in place for staff to aid in the event a subject access request is sent to them. Finally, it is general good practice to have a policy for any access request made and, to formally document and record all access requests that have been made by data subjects of the organisation.

It would also be beneficial, for the organisation, to have within the policy a standardised document for DSAR's, within the organisation. Recital 59 of the GDPR states that organisations should 'provide means for requests to be made electronically, especially where personal data are processed electronically'. This should be incorporated into organisational policy as stated previously, designing a subject access form that facilitates ease of use, for both the individual and the organisation. Although it should be noted that even if the organisation has a standardised form their internal policy, any DSAR can be made without using the form provided and must still be granted under the requirements of the GDPR.

Chapter Review

Summary of Key Points

1) DPO: Appointing a DPO.
2) DPO: Tasks of a DPO.
3) DPO: Qualities of a good DPO.
4) DPIA: Legal Requirements.
5) DPIA: Defining Article 35.
6) Conducting a DPIA.
7) DSAR: The role of the organisation.
8) DSAR: How to access your data.

Cognition

- Consider how an organisation should handle a DSAR.
- Think about what qualities you would like to find in a DPO and how they relate to data protection.
- Think about the kinds of risks that would arise in a project involving vulnerable members of society if you were to conduct a DPIA.

References

Calder, A. (2018). *Data protection impact assessments under the GDPR*. Retrieved from ITGovernance: https://www.itgovernance.co.uk/privacy-impact-assessment-pia

Data Protection and Privacy Commissioners. (2010). Resolution on Privacy by Design. *ICDPPC*, (pp. 1-2).

European Data Protection Supervisor. (2015). Towards a new digital ethics. *Opinions*(4/2015).

Farber, D. (2018). Everything You Need to Know about the Data Protection Officer Role. Belgium: HackerOne.

ICO (Information Commissioner's Office). (2014). Conducting privacy impact assessments code of practice. *Ico.Org.Uk*.

ICO. (2015). Information Security (Principle 7). In ICO, *Guide to data Protection, Information Commissioner's Office (ICO)*.

ICO. (2015). Key definitions of the Data Protection Act. *Internet*.

ICO. (2015). The Guide to Data Protection. *Information Commissioner's Office*.

IT Governance. (2016). *The data protection officer (DPO) role under the GDPR*. Retrieved from it governance: https://www.itgovernance.co.uk/data-protection-officer-dpo-under-the-gdpr

Massey, S. (2017). *The Ultimate GDPR Practitioner Guide*. London: Fox Red Risk Publishing.

The European Parliament, & The European Council. (2016). General Data Protection Regulation. *Official Journal of the European Union*.

Tikkinen-Piri, C., Rohunen, A., & Markkula, J. (2018). EU General Data Protection Regulation: Changes and implications for personal data collecting companies. *Computer Law and Security Review*.

UK Information Commissioners Office. (2017). Subject access request. *Guide to data protection*.

Appendix

Appendix 7a: DPIA Template

Step 1: Identify Need

Here the organisation should briefly explain the need for the DPIA: naming things such as the project, technologies, and processes involved and what the project aims to achieve.

Step 2: Describe the Processing

Describe the Nature of the Processing: how the collection, storage, and deletion of the data will occur (data flow map)	*Describe the Scope of the Processing:* Nature of the data, does it include special categories of data etc. how many data subjects will be affected	*Describe the Context of Processing:* How much control will DS have, does the processing involve vulnerable groups (children), is the org. using a code of conduct	*Describe the Purpose of Processing:* What is the effects on the individual, what will the project achieve, what are the benefits of processing

Step 3: Consultation

Article 36 requires prior consultation for high risk projects: here the Org should consider how to consult with the relevant stakeholders. Should the org consult experts or the SA? How will the org seek out the individuals affect and when?

Step 4: Asses Necessity and Proportionality

What is the lawful basis for processing for the new project? Are there alternatives to achieving the same outcome? How will the project fulfil the principles of the GDPR, such as data minimisation? How will the project uphold the rights and freedoms of the natural person?

Step 5: Identify and Assess Risk

Identify Risk: Here the Org should identify what the potential risks are to the operation, the source and impact if it were to occur	*Likelihood of Harm:* After identifying the risk, the org should then assess the likelihood of that outcome some useful terms: Highly Unlikely, Possible, or Highly Likely. As an ex	*Severity of Outcome:* Here the org should asses the level of harm in the event the risk was to be realised (minimal damage, significant, or severe).	*Overall Risk:* After evaluating the likelihood and severity this section should evaluate the overall risk, Ex. High likely hood minimal damage = low to medium risk.

Step 6: Identify Measures to Mitigate Risk			
Risk: Like step 5, the named risk should go here	*Measures to Reduce or Eliminate Risk:* Here the org should find solutions to mitigate the risk to the privacy of the individual	*Effects on Risk:* If the measure were to be carried out how would it affect the risk. Ex. Risk of data breach, solution encryption = reduced risk	*Residual Risk:* Here after evaluating the effect the measure had on the risk what is left over should be mentioned here. Previous ex = from high risk to low risk.

Step 7: Sign off and Document Outcomes
Here the DPO should write off on any measure taken and either approve or refuse them, with some commentary on what can be done better or to advise on company policy. In this section the names of any individuals involved should be documented and stored for review by the SA if needed.

Appendix 7b: DSAR Letter Template

The ICO has published a recommended letter template for data subjects to exercise their right of access the letter is displayed below:

[Your full address]
[Phone number]
[The date]

[Name and address of the organisation]
Dear Sir or Madam
Subject access request

[Your full name and address and any other details to help identify you and the data you want.]

Please supply the data about me that I am entitled to under data protection law relating to: [give specific details of the data you want, for example:

[• my personnel file
• emails between 'person A' and 'person B' (from 1 June 2017 to 1 Sept 2017)

- my medical records (between 2014 and 2017) held by 'Dr C' at 'hospital D'
- CCTV camera situated at ('location E') on 23 May 2017 between 11am and 5pm
- copies of statements (between 2013 and 2017) held in account number xxxxx.]

If you need any more data from me, or a fee, please let me know as soon as possible. It may be helpful for you to know that data protection law requires you to respond to a request for data within one calendar month.

If you do not normally deal with these requests, please pass this letter to your DataProtection Officer, or relevant staff member. If you need advice on dealing with this request, the Information Commissioner's Office can assist you. Its website is ico.org.uk or it can be contacted on 0303 123 1113.

Yours faithfully

[Signature]

3

Part Three: Implementation

Chapters Covered

8) International Standards; ISO's
9) Security Incident Management
10) Valuing Security

Section Overview

This section aims to address the areas of implementation. These few chapters are fundamental to understanding the GDPR in action. It begins with international standards of risk and other key areas of management and there relations with the GDPR, followed by actions taken by organisations when things go wrong, and finally how to make the business case for security to senior management.

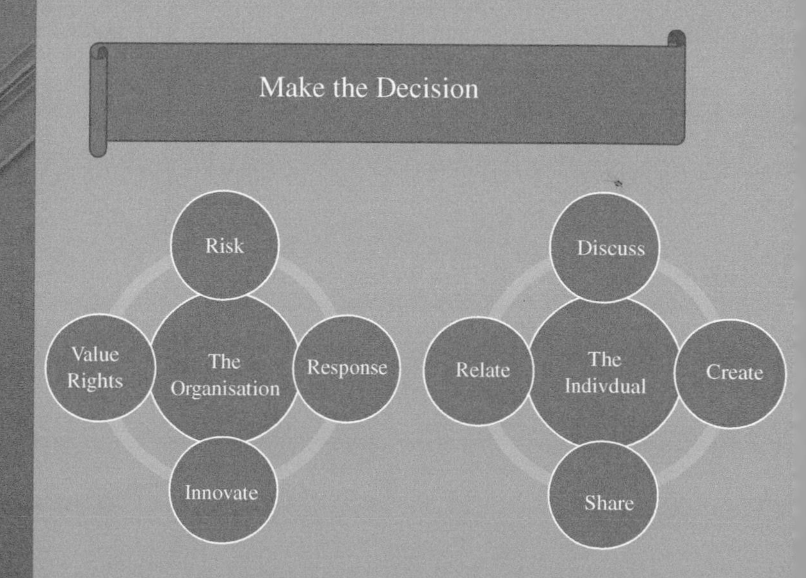

Chapter 8: International Standards; ISO's

At a Glance:

- ISO 31000: Risk Management
  - The Eight Principles
  - Five Component Framework
- ISO 8601: Time Formatting
- ISO 27000: ISMS
- ISO 9001: Quality Management

Learning Objectives: *Students should be able to…*

- Understand the role of industry standards in regulatory compliance.
- Describe the process for the creation of an ISO standard.
- Explain the scope of application in the included standards.
- Describe the key concepts in the included standards.

Key Terms

1) International Standardization Organization: ISO
2) Risk Management: RM
3) Information Security Management Systems: ISMS
4) Date and Time standardisation
5) Quality Management: QM
6) Plan, Do, Check, Act Cycle: PDCA

The ISO

This chapter presents an overview of relevant ISO standards for Business Information Systems; as an aid in the compliance to the GDPR and in implementing best practises for cyber security.

The International Organisation for Standardization is known by the acronym, ISO. It is a federation of national standards bodies from across the world. The ISO develops International Standards through working groups called ISO Technical Committees, and any member body wishing to participate in the work of a specific Technical Committee, has the right to do so. Member bodies represent relevant industries, academia, government bodies, consumer associations, and non-government organisations. The ISO also works in close collaboration with the International Electrotechnical Commission (IEC) wherever standardisation of electrotechnical issues are required.

The stakeholders (member bodies) are drawn from a wide range of interested parties, including industry experts nominated by the member bodies. This broad level of input provides the authority and relevance to the use and application of the ISO standards, internationally.

4 Key Principles

1) *Responsive:* New standards are developed as a response to requests from stakeholders; industry or consumer groups for example. These requests are made to the ISO member of the relevant nation who then raises it with the ISO.

2) *Expert Opinion:* The details of the standards are developed by industry experts globally. These experts explore and analyse all aspects of the proposed standard. They then inform, and are part of, the ISO Technical Committee relevant to their area of expertise.

3) *Multi-stakeholder process:* Whilst the opinion of industry experts is a key principle in developing standards, the input from other stakeholders is important in giving the standard legitimacy. Representation on the Technical Committee of a wide range of stakeholders, provides valuable information on the opinions of groups who will be impacted by the standards. Stakeholders are those people and organisations can be, or believe themselves to be, or are, affected by the decisions and activities of the organisation.

4) *Consensus:* A consensus-based approach means that all stakeholders are listened to, and their input is of equal importance. Again, this is fundamental to the notion that legitimacy, and acceptance of use, is conveyed through an open and democratic approach to the development of international standards.

5 Year Review Process

Every ISO standard is reviewed every 5 years to ascertain the need, or not, for a revision. Standards must be kept current and relevant for their continued use and especially in fast changing industries, such as cyber security. Revisions respond to the most recent trends and changes affecting the standard under review, and this process encourages users of the standards to feel confident in its application.

ISO as a Function of Compliance

In the following pages the ISO Standards which are helpful in the application of the Principles within the GDPR will be outlined. These are not exhaustive but serve as an introduction to the Standards often mentioned in the context of data protection. As these standards are internationally accepted, they form a solid foundation for the "best practice" approach to data protection. Part of this best practice is an understanding of broader management principles; especially Risk Management, since the GDPR emphasises a risk-based approach to compliance.

ISO 31000: Risk Management

'the effect of uncertainty on objectives'

Risk is the uncertainty involved in the achieving of desired outcomes. This uncertainty arises from both internal and external factors for an individual or an organisation, as all activities involve some risk. Risk is not, in itself, a negative factor for without taking risks one may not maximise opportunities, however it is in mitigating *undesired* outcomes that risk management is especially powerful. Organisations achieve this through the process of:

- Identifying risk;
- Analysing risk factors;
- Evaluating the level of risk acceptable.

Risk treatment allows the level of risk inherent in a process or activity to be modified within acceptable parameters in order to meet the organisation's risk appetite; that is, the potential impact on the desired outcome, or goals. Throughout the period of risk exposure, communication with stakeholders and monitoring of risk factors allows organisations to modify the approaches for minimising risk. ISO 31000 is the international standard for the systematic and logical process in managing risk. It entails set principles which increase the effectiveness of risk management frameworks by integrating them into the whole organisation; at an overarching level such as governance, and at the detailed level of specific activities or projects.

This standard can be used by organisations of any size or industry sector, as the application of these practises relate to the understanding and mitigation of risk, and not to the specific requirements of an industry. It provides a consistent, credible, and transparent framework which allows the organisation to achieve its goals and to meet certain regulatory requirements in its documentation outputs. The broad application of this framework requires the specific context to be delineated, for each sector or risk scenario.

Risk management refers to the overall structure or architecture for the effective management of risk; the principles, framework, and processes. There are eight principles, six framework components, and six processes within this standard. All of which are intended to satisfy the needs of a range of stakeholders, especially those involved with the development and management of risk policies and procedures within an organisation.

The Eight Principles

There are eight principles to risk management that act as the basis to executing a well-developed risk strategy. Below is a brief introduction to the Eight Principles of risk management:

1) *Integrated:* At the operational and strategic level risk management should be an integral part of organisational processes and integrated with the governance framework.
2) *Structured and Comprehensive*: In order to ensure efficiency, consistency and the reliability of results. Across the organisation the process must be consistent, systematic, structured and timely.

3) *Customised:* The risk management framework of the organisation needs to be tailored to the scope, context, and criteria of its risk profile, and take into consideration the internal and external operating environment.
4) *Inclusive:* Be transparent and inclusive by engaging stakeholders, (internal and external), throughout the risk management process. Recognise that communication and consultation with stakeholders is key to identifying, analysing and monitoring risks.
5) *Dynamic:* The process of managing risk needs to be flexible; a dynamic, iterative and responsive approach is most effective.
6) *Best available information:* It is important to understand and consider the relevance of the best available information, and to be aware of limitations to that information. Understanding the applicability of the information is also important.
7) *Human and cultural factors:* Consider the human and cultural factors means that risk management needs to recognise the contribution that people and the culture have on achieving the organisational objectives.
8) *Continual improvement:* Facilitate the continual improvement of the organisation's risk management culture; by learning from experience.

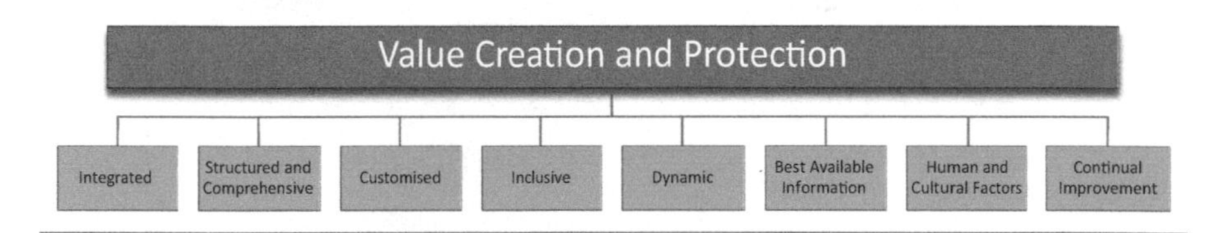

Five Component Framework

The framework for risk management is defined by the ISO as:

"a set of components that provide the foundations and organizational arrangements for designing, implementing, monitoring, reviewing and continually improving risk management throughout the organization".

The 2018 revised framework has the following components:

- *Leadership and commitment:* Stakeholders must be committed to supporting the management of risk. The top management and oversight bodies must be identified as leaders of the integration process; which they oversee and are held accountable for.

- *Integration:* risk management must be part of the whole organisational system; from governance and strategy through to operations.
- *Design:* designing risk management into every aspect and activity undertaken.
- *Implementation:* policies, procedures, and activities must be systematically applied in a coherent and workable way.
- *Evaluation:* ongoing monitoring of the framework relative to the goals of the organisation. Regular review of the framework's effectiveness and relevance to the developing needs of the organisation.
- *Improvement:* continual improvement of the framework based on its suitability, effectiveness, and the understanding of how to adequately provide risk management.

Six Stage Process

Communication and consultation:

Maintaining an open dialogue with stakeholders about the management of risk in the organisation is a key part of the process. The aim of communication is to share information with, and provide information to, people and groups. Consultation is the process which impacts decision making through its input and influence. It is the *informed* two-way communication which occurs between relevant parties, usually an organisation and its stakeholders, *before* any decision-making process is undertaken.

Scope, context, criteria:

The first step in the risk management (RM) process is to establish the context so that the nature and complexity of the risks may be understood. This activity details the organisation's objectives and asks questions about the nature and scope of the factors influencing the outcome. These questions aim to gather information on:

1) The environment in which the organisation seeks to achieve its objectives.
2) The stakeholders involved.
3) The diversity and range of the risk criteria.

Risk Assessment:

The overall process of assessing risk is made up of three parts:

1) *Identification:* the process of finding and recognising the sources of risk and describing the potential causes and consequences of risks.
2) *Analysis:* is the basis on which risk is evaluated and decisions about risk treatment and risk minimisation are made. It is the process used for understanding the nature of the relevant risk and allows the level of risk to be estimated.
3) *Evaluation:* provides information which helps to assess the nature and efficacy of risk treatments. It is the process whereby the results of the risk analysis are used to decide if the risk is acceptable to the organisation.

Risk treatment:

Risk treatment is the process used to modify exposure to risk. This process deals with both risks with negative impact, or risks with potentially positive impacts. Avoiding the activity or removing the source of risk are means that may be used to deal with risks which are assessed as having a negative impact on the desired outcomes of the organisation. This type of treatment can also be referred to as: mitigation, prevention, elimination, or reduction of risk. On the other hand, an organisation may choose to *increase* the risk to exploit an opportunity, or they may share the risk exposure with another organisation.

Monitoring and review:

Monitoring is the need to continually check and supervise the performance of the risk framework by critically observing any change from that needed or expected. Review is that process where the effectiveness and suitability of the framework being used is measured against the established objectives.

Recording and reporting:

The process wherein the above activities and their outcomes are communicated to stakeholders and across the organisation. It must be structured in such a way so that activities and decision-making can be improved.

ISO 27005: A Brief Visit

ISO 27005: Information security risk management was last updated in 2011 and has been under review since 2015. Due to the lack of clarity surrounding the current state of this standard, the ISO 31000 has been detailed instead since it is applicable to the management of risk in any industry and environment. ISO 31000 is widely used throughout the world as a best practice standard and has been updated in 2018.

ISO 8601: Representation of Dates and Times

ISO 8601 is the international standard for the exchange of date and time related data. It is used for the writing of the representation of times and dates in an unambiguous and clearly defined manner. This is to avoid the potential confusion which can arise in the understanding of date and time representations, especially when data is transferred between countries.

When dates and times are represented using numbers it is important that they are able to be interpreted correctly. For example, 06/03/2018 may be interpreted as the 6th of March or the 3rd of June, depending on the reader's usual dating format. This has obvious implications in the smooth and efficient running of all organisations. To remove this uncertainty, the ISO 8601 sets out the internationally accepted format for the numeric representation of dates and time. Dates and times are written in terms of increasing accuracy; the broadest term first, followed by successively more specific ones. The largest temporal term is written first, and each successive value is placed to the right of the largest one.

Times are therefore written with the hours, then minutes, then seconds: HH: MM: SS.

- Twenty minutes past four in the afternoon would be represented as; 16:20:00 in the twenty-four-hour timekeeping system, and as 04:20:00 in the diurnal system.

Dates are written with the year first, then the month, and then the day: YYYY-MM-DD.

- The 25th May 2018 would be represented as; 2018-05-25.

The standard applies to the Gregorian calendar, the 24-hour timekeeping system, time zone information, and time intervals. It cannot accommodate systems which require words or characters to be included; with the exception of those letters, words, or certain characters which have a specific numerical value assigned to them. Applying this standardised date and time format across an organisation will increase efficiency, reduce confusion and ambiguity, and harmonise date and time data internationally.

Many organisations and government bodies use ISO 8601 as their standard for date and time. All UK Government departments and bodies are mandated to use this system. Organisations that value precision, such as NASA, also use it, as modern programming languages support this standard. It is considered especially useful because it is easy for people to understand, whilst providing the high precision required for machine-to-machine communication.

Whilst ISO 8601 does give increased precision, organisations involved in activities requiring people to engage in specific behaviours, at set times, may also need to highlight days and times in words. For example, the NHS has guidelines for the writing of medicine labels which emphasise the *name* of the day on which the medicine is to be taken.

ISO 27000 Family – Information Security Management Systems

This family of international standards is clearly relevant to Information Systems, and as such plays an important role within compliance to the GDPR. This standard helps organisations to manage the security of information systems and is presented in over 45 sub-sections. Information Security Management Systems (ISMS) aim to secure sensitive information within organisations through the management of risk processes as applied to; people, processes, and IT systems. Any organisation, regardless of size, may benefit from the application of the methods encompassed in ISMS. This standard seeks to enhance, through the application of risk management processes, the confidence of stakeholders in the secure management of information. This is provided through the preservation of information through:

- *Confidentiality:* information is not made available or disclosed to unauthorised individuals, entities, or processes.
- *Integrity:* accuracy and completeness of data and information.
- *Availability:* information is available on demand and is accessible and usable to those authorised to access it.

ISMS: Information Security Management Systems

Information security management encourages organisations to treat information in a similar way to other business assets through approaching ISMS as an ongoing project. The framework is made up of policies and procedures covering areas of control involved in the process of risk management within an organisation.

Data protection by design and default is implicit within ISO 27000, as it places great importance on the integration of ISMS with organisational structure and processes. Information security must be considered in the design of processes throughout an organisation, and within information systems. Organisations are required to assess and treat information security risks in a manner that is specific to their needs. ISO 27001 is considered to be best practice for an ISMS and provides a framework for the secure management and protection of information. Certification is available to those organisations who fulfil the requirements of the standard.

An audit of the organisation's ISMS must be performed by an accredited certification body; and where successful, the organisation will achieve accreditation.

ISO 27001 certification is a rigorous and costly standard to achieve, however its benefits are numerous. As an internationally recognised and used set of standards, it allows organisations to harmonise their ISMS across multiple sites, and with other organisations. Although an ISMS must be tailored to the specific needs of each organisation, the standardisation of the frameworks used within each organisation creates a "baseline" from which all stakeholders may operate.

Although this standard directly relates to only one of the Articles (32) in the GDPR, its emphasis on documentation of the processes and procedures used, and its iterative function, provide a strong foundation. Certified organisations are regularly monitored through external audit for compliance with the standard, and the awareness of this ongoing scrutiny can be a useful primer for any potential audit by the Supervisory Body for the GDPR. The GDPR emphasises a risk-based approach to data security, and so does this standard, thus providing a useful tool in achieving compliance.

The standard provides the requirements for ISMS to establish and implement the system: informed by the needs and objectives of the organisation, its structure and size, the processes used, and its security requirements. These factors, which influence the establishment and implementation of ISMS may, and likely will, change over time. The requirements also provide details on how to maintain and continually improve the system over time, and can be used to assess an organisation internally, or by external third parties, such as a regulator. It applies a top-down, risk-based approach implemented through a planning process in six parts, these are:

1) Define a security policy
2) Define the scope of the ISMS
3) Perform a risk assessment
4) Manage the risks identified
5) Select the control objectives, and the controls to be implemented
6) Prepare a Statement of Applicability (SoA)

ISO 27018: Code of Practice for Protection of Personally Identifiable Information (PII) in Public Clouds Acting as PII Processors

This standard has been developed for the protection of Personally Identifiable Information (PII) as applied to services in the Cloud. It directly relates to information security risk environments pertinent to providers of public cloud services. ISO 27018 is suitable for organisations of all sizes, and of any industry sector; including government bodies, and not-for-profit organisation; where they provide PII processing services through cloud computing under contract to third parties. It is especially useful and relevant where organisations are already protecting their information assets; for example, by using ISO 27001.

Cloud service providers provide processing services, under contract to their customers. Where this processing involves the processing of PII, they must operate in a way that meets the legal requirements of *both* their own organisation and that of their customers, regarding data protection. The requirements vary depending on the jurisdiction (legal zone) of each party, and the contractual agreements between them. The varying requirements in differing jurisdictions creates a challenging environment for compliance and the ISO 27018 aims to harmonise standards across jurisdictions.

The main objectives of this standard are to help cloud service providers, when acting as a processor of PII, to comply with regulatory obligations. This is achieved by assisting the cloud service provider to enter into contracts with customers based on a well-governed and transparent system. This also allows the customer to easily interact with the system, to audit what is happening to their PII. It establishes control objectives, which are widely used, and provides controls and guidelines for the implementation of PII security measures.

The need for PII security may arise from three main types of requirement:

- *Legal:* these may be statutory, as in legislation, regulatory, or contractual.
- *Risks:* the risks the organisation may be exposed to associated with PII.
- *Policies:* organisations may decide to secure PII beyond the levels indicated by the legal and risk requirements.

The choice of which controls to apply is specific to the needs, and general attitude to risk, of the particular organisation. The ISO 20018 provides a selection of controls, from which the organisation should choose. The type of controls chosen will also be dependent on the actual role the organisation fulfils within the cloud computing architecture, as a whole.

ISO 27032: Guidelines for Cybersecurity

ISO 27032 provides guidelines for safe interaction between organisations and third parties and processes in cyberspace. The definition of cyberspace, as outlined in this standard is the:

> *"... complex environment resulting from the interaction of people, software and services on the Internet, by means of technology devices and networks connected to it, which does not exist in any physical form."*

Cybersecurity is the practice of applying security measures for the protection of interactions within cyberspace through the preservation of confidentiality, integrity, and availability. ISO 27032 provides an overview of cybersecurity and its relationship to other security disciplines. Stakeholders are defined and their roles relevant to cybersecurity are described along with a framework for collaboration in the addressing and resolution of cybersecurity issues. Risks common to the cyberspace are addressed and technical guidance for mitigation of these risks is outlined and controls for the prevention, detection, and response to cyber-attacks are included.

Risks are identified relative to the organisation's assets present within the cyberspace, these are:

- Information
- Software
- Physical
- Services
- People
- Reputation

ISO 27032 also provides a framework for the secure and reliable exchange and sharing of information between stakeholders, which enables them to coordinate activities, and effectively handle incident response. The framework contains elements upon which trust may be built, processes for information sharing, exchange, and collaboration, and technical requirements relevant to systems integration and functionality between stakeholders.

The framework consists of components necessary in addressing common cybersecurity issues between stakeholders; the information providing organisation and the information receiving organisation. This is achieved through the following components:

Policies:

- ❖ Define cybersecurity incident life-cycle
- ❖ Classify and categorise information
- ❖ Minimize information
- ❖ Protocol for coordination activities

Methods and processes:

- ❖ Use non-disclosure agreements (NDA)
- ❖ Code of practice
- ❖ Scheduled sharing of information
- ❖ Monitoring and testing

People and organisations:

- ❖ Key stakeholder contact details
- ❖ Internal and external industry alliances
- ❖ Staff awareness and training

Technical:

- ❖ Standardisation of data
- ❖ Data encryption and key exchange
- ❖ Backup software and hardware
- ❖ End-to-end secure file sharing and communication
- ❖ Systems testing

ISO 9001 – Quality Management System

The ISO 9001 is a Quality Management System (QMS) that is a useful methodology due to its relevance across all industries and the fact that it is the only standard in this family which has certification available. ISO 9001 can be used by any organisation, regardless of size or industry sector, and is based on the concept of continual improvement.

This standard sets out how organisations may ensure consistent provision of good quality products and services to their customers. This is based on QMS principles such as; customer focus, motivation of top management, the Process approach, and continual improvement. Quality Management System (QMS) is described as the principles (policies) and processes (procedures) involved in the design, development, and delivery of products and services. ISO 9001 certification is sought by organisations who wish to gain the benefits arising from a coherent and consistent internal framework in achieving customer focussed goals and want their customers to know about it. Certification is awarded to organisations who meet the specific requirements outlined.

The standard is applicable to organisations who need to demonstrate their ability to consistently:

- ❖ Meet customer requirements.
- ❖ Comply with regulatory requirements.
- ❖ Enhance customer satisfaction.

Risk-based thinking

Risk-based thinking allows organisations to prioritise what is acceptable, and what is not, when assessing risks and opportunities. Its primary objective is the prevention of undesired outcomes; whether it effects products or services. ISO 9001 requires organisations to plan for, and implement, preventative measures to minimise or eliminate potential risks, and where undesired outcomes have occurred, action must be taken to prevent any reoccurrence. A secondary, but important objective, is the ability to maximise opportunities which allow the organisation to achieve its goals. Opportunities may be identified through the analysis of risks, highlighting areas which may have been previously overlooked.

The benefits of using a Quality Management System include:

- ❖ The ability to provide products and services consistently.
- ❖ Opportunities for enhancing customer satisfaction.
- ❖ Address risks and opportunities.
- ❖ Demonstrates conformity and compliance.

To achieve these objectives the Process approach is used, which assists in the planning of processes and interactions. One of the tools used is the PDCA; Plan, Do, Check, Act cycle, which helps organisations to adequately resource and manage their processes. It also helps to identify areas for improvement which should be addressed. Another aspect employed in the Process approach is Risk-based thinking. Primarily used as a preventative tool, it helps organisations to identify factors that could lead to deviations from the desired goal within the system or process. Risk-based thinking aims to minimise the negative impacts of undesirable factors, and to maximise the benefits of opportunities that appear.

The Process approach

The use of the Process approach in the development and implementation of a Quality Management System (QMS) is encouraged to enhance the effective improvement of the system. The Process approach allows organisations and managers to systematically define processes and the level of inter-related activities and dependencies among those processes. Understanding what processes are active within the system provides the opportunity to analyse their effectiveness relative to the organisations' goals. The management of those processes can then be adapted or changed to enhance the overall effectiveness of the system, and to increase efficiency. The benefits of a Process approach are:

- Increased understanding and improved consistency
- Added value through effective processes
- Information based improvements

Plan-Do-Check-Act Cycle

Used as a method for the continual improvement and control of processes, the PDCA cycle can be applied to a specific process or to a system as a whole. One of its strengths is that it is easy to apply, having only four steps, which are repeated, hence the cycle.

The four steps are:

Plan

- Establish objectives of: the system; and of its processes
- Ascertain the resources required to meet: customer requirements; and organisational policies

- Identify: risks; and opportunities
- Address: risks; and opportunities

Do

- Implement the plan

Check

- Monitor: processes; products; and services
- Measure: processes; products; and services
- Report: on the results relative to objectives

Act

- If the report shows an improvement from the original (baseline) system, then the Plan becomes the new baseline standard. It becomes the new "Act", that is the new process.

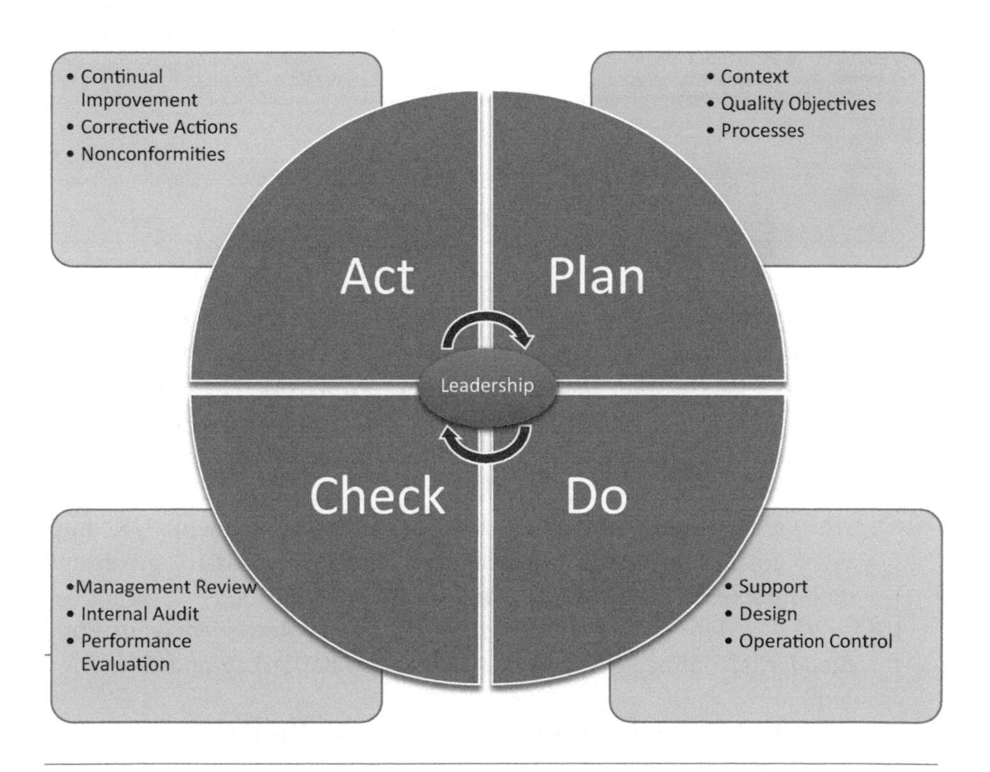

Chapter Review

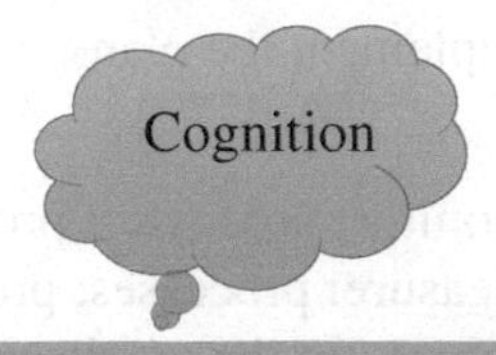

Summary of Key Points

1) Nature and function of the ISO.
2) RM is integral to GDPR compliance.
3) ISMS provide structure.
4) Need for Date and Time accuracy.
5) QM contributes documentation.

- Which ISO standards would you use?
- Could you apply the ISO concepts to everyday life?
- How would you know which ISO standard is best for an organisation?
- What changes could be introduced to your organisation, based on ISO ideas?

References

American Society for Quality. (2018). *ISO 9000 Quality Management Principles*. Retrieved from Learn About Quality: http://asq.org/learn-about-quality/iso-9000/overview/quality-management-principles.html

Deming, W. (1960). *Some theory of sampling*. New York: Dover Publications.

Dutton, J. (2018). How ISO 27001 can help to achieve GDPR compliance. London: IT Governance UK.

Government Digital Service. (2018). *Guidance: Date-times and time-stamps*. Retrieved from Open Standards for Government: UK: https://www.gov.uk/government/publications/open-standards-for-government/date-times-and-time-stamps-standard

ISO. (2004). *Date and time format – ISO 8601*. Retrieved from Popular Standards; online: https://www.iso.org/iso-8601-date-and-time-format.html

ISO. (2009). *ISO Guide 73: Risk management — Vocabulary.* International Standards Organisation, Geneva.

ISO. (2011). *ISO/IEC 27005:2011*. Retrieved from Standards Catalogue; online: https://www.iso.org/standard/56742.html

ISO. (2012). *ISO/IEC 27032:2012*. Retrieved from Standards Catalogue; online.

ISO. (2013). *ISO/IEC 27001:2013 I*. Retrieved from Standards Catalogue; online.

ISO. (2015). *ISO 9001:2015*. Retrieved from ISO 9000 family – Quality management: https://www.iso.org/iso-9001-quality-management.html

ISO. (2018). *About ISO*. Retrieved from All about ISO: https://www.iso.org/about-us.html

ISO. (2018). *ISO 31000:2018 Risk management – Guidelines*. Retrieved from ISO.org: https://www.iso.org/standard/65694.html

ISO. (2018). *ISO 9001:2015*. Retrieved from ISO 9000 family – Quality management: https://www.iso.org/iso-9001-quality-management.html

ISO. (2018). *ISO/IEC 27000:2018 I*. Retrieved from Overview and vocabulary: https://www.iso.org/standard/73906.html

ISO. (2018). *ISO/IEC 27018:2014*. Retrieved from Standards Catalogue; online: https://www.iso.org/standard/61498.html

Walton, M. (1986). *The Deming Management Method*. London: The Putnam Publishing Group.

ISO. [illegible] *ISO/IEC 27005:2011*. [illegible] from Standards Catalogue [illegible] https://www.iso.org/standard/[illegible]html

[illegible] *ISO/IEC 2[illegible]:2012*. Retrieved [illegible] Standards Catalogue [illegible] online.

ISO. (2013). *ISO/IEC 27001:2013* [illegible] online.

ISO. (2015). *ISO* [illegible] from ISO 9000 [illegible] [illegible]-9001-quality-management [illegible]

[illegible] about-us [illegible] https://www.iso.org/ [illegible]

ISO. (2018). [illegible] Retrieved [illegible] https://www.iso.org/standards.html

ISO. (2017). [illegible] Retrieved from ISO 9000 family – Quality [illegible] iso-9001-quality-management [illegible]

[illegible]

[illegible] Standards [illegible] Retrieved [illegible] standard/[illegible].html

[illegible]

Chapter 9: Security Incident Management

At a Glance:

- Introduction
  - o CSIRT and the articles of the GDPR
- Incident Response Plan
  - o Incident Life Cycle: Detect Respond Analyse
- Calculating data breach severity using the ENISA method
  - o How it affects CIA

Learning Objectives: *Students should be able to…*

- Identify the importance of a CSIRT team.
- Understand the articles relating to data breach notification and when to notify the Supervisory Authority.
- Explain the elements of the incident response plan and the OODA Loop.
- Explain how to calculate data breach severity using the ENISA methodology.

Key Terms

1) Computer Security Incident Response Team
2) Data Breach
3) Notification
4) Data Breach Severity
5) SIEM
6) Incident Life Cycle

Introduction

Data breaches have and continue to be an ongoing struggle in an information sensitive world. Breach Level Index has indicated that since 2013 over 9 billion data records have been lost or stolen, where only 4% of cases the data breaches were considered "secure breaches" or breaches where encryption was used. Most breaches can be attributed to poor internal security. Stolen or lost records caused by data breaches can also cause financial loss to an organisation, with Oxford Economics finding that out of a sample of 65 companies, data breaches caused around $52 billion cost to shareholders.

A data breach in the context of the GDPR is the accidental or unlawful destruction, loss, alteration unauthorized disclosure, or loss of availability of personal data. This differs slightly from a general computer security incident, in that, any attempt (successful or not) to gain unauthorized access to personal data or the system itself, denial of services, and unwanted changes to system hardware, software etc is considered a security incident.

Incidence response is an organized approach when dealing with the effects of a security breach or cyberattack. Responding to an incident of a security breach or cyberattack should be done in a way that minimizes material, non-material, or physical damage to the data subject.

There are different methods in dealing with data breaches, this chapter will look at the relevant parts of the GDPR relating to incident response, preventative steps an organisation can take, what to do in the event of a data breach, the severity of a data breach, and the notification to the relevant parties affected by a data breach.

The GDPR Articles

Under the GDPR, incident response, relates to Articles 31–34.

Compliance to Article 32 determines the level of security which was applied before responding to an incident of a data breach. Summarized, depending on the type of data that is being processed appropriate measures should be taken to protect that data (see chapter on cyber security and privacy by design 4 and 5). This article is more of a preventative measure required by the regulation that can assist the organisation and the Supervisory Authority in the event a breach does occur. As discussed in chapter 5 protection by design is the foundation of Article 32, and is an effective way an organisation can best avoid a breach from occurring, given data protection is built into the corporate culture. Although protection and privacy by design and default is necessary in creating the culture it doesn't make the organisation immune from attacks.

The primary articles relating to incident response in the GDPR are Articles 33 and 34. Article 33 refers to the notification of a personal data breach to the Supervisory Authority, whilst Article 34 is the notification to the data subjects in situations the breach is likely to cause high risks to the rights and freedoms of the data subjects. The main points that should be noted in Article 33 is that notification should be reported no later than 72 hours after having become aware of a data breach, unless the breach is unlikely to infringe on the rights and freedoms of the natural persons (outline in Article 33(1)).

In the context of an organisation the 72-hour deadline is the most important point, the prime question being what to do in that 72 hours?

Computer Security Incident Response Team (CSIRT)

The first things to consider when a data breach occurs is to be prepared. Incidences that are not properly contained can quickly escalate into greater problems that can result in a damaging breach for the personal data subject and/or a complete system collapse for the organisation.

The Computer Security Incident Response Team, CSIRT (pronounced see-sirt), is an organisational body that deals with the on-going effects and aftermath of a security breach, but can also aid in preventing data breaches from occurring. The CSIRT can either be integrated into the business as an established entity or an ad hoc group. Typically, larger organisation will have a subsidiary company or separate department (like a C-suite) that operates as a CSIRT, known as an internal CIRST, as opposed to external CSIRTS which may deal in handling data breaches for an entire country or as a service (third party).

The primary difference between the two is, an internal CSIRT could be made up of organisational personnel that would typical hold other jobs in the day to day activities of the organisation, while an external CSIRT is a dedicated service dealing with data breach incidence response either as an on-going service or on a "when-needed" basis. One example of this is Japan Computer Emergency Response Team Coordination Centre (JPCERT), an external CSIRT, which deals with computer security incidences on a country wide basis. The CSIRT should follow a conventional organisational hierarchy, comprised of key organisational personnel. Depending on the size of the organisation some companies may not have the resources to completely fill out the CSIRT team as shown in figure 9.1, but should strive to have at least the top half if they are processing sensitive personal data.

For the GDPR it is crucial that the CSIRT of a highly technology dependant organisation involves both the Data Controller and the Data Processor. In that the Controller involves all the relevant Processors that they are collaborating with, and that the Processors incorporates relevant members from the Controllers organisation into their respective CSIRT teams. This can provide greater preparation in the event of a data breach and facilitate faster communication between the two organisations.

Figure 9.1: CSIRT Hierarchy

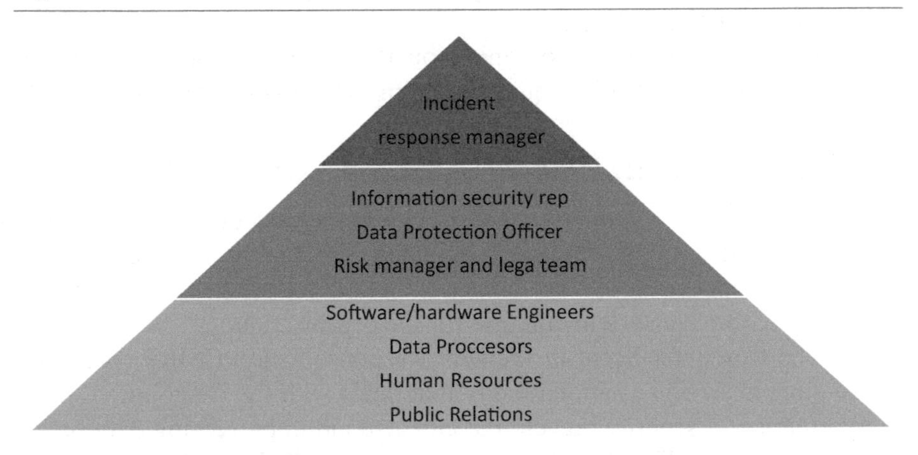

This diagram shows an example of the organisational structure of a CSIRT, the top half are strongly recommended components for the team, while the bottom segment is a welcome, but not necessary, addition.

In the global economy of information, it is vital that organisations and government bodies have a readiness plan when dealing with personal data breaches, forming a CSIRT team is a great way of showing compliance with the regulation and that the organisation works toward protection by design and default. Recital 87 dictates that:

"It should be ascertained whether all appropriate technological protection and organisational measures have been implemented to establish immediately whether a personal data breach has taken place..."

The CSIRT would be the department to ascertain that appropriate measures were taken and that the response to the incident is carried out in a timely manner with least damage done to the data subjects.

Incidence Response Plan (IRP)

The IRP is the protocol, or a risk management plan, put in place by an organisation that the CSIRT can execute in the event of a personal data breach. The protocol should comply with the relevant parts of the GDPR, aforementioned in the previous sections (Articles 33, Article 34). This is essentially what the organisation should be doing in the 72 hours after having become aware of the breach. The role of the CSIRT when executing the IRP can be broken down into three main elements recognize, analyse, and respond (which will be fully elaborated in the incident response life cycle section).

The first step a CSIRT should undertake is to recognize that a data breach is even taking place. This can be achieved through various technological capabilities (such as intrusion detection systems IDS), and ascertaining if the security incident is malicious or not. There are other ways a personal data breach can occur, and this can cause challenges in recognition, the most obvious example is if data is lost or misplaced by organisational staff or a malicious insider who wishes to use his or her admin capabilities to gain unauthorized access into the system, then it becomes a matter of *when* a data breach occurs rather than if a breach will occur.

This becomes a managerial issue over a technological one, as the Breach Level Index report showed that in the first half of 2017 over 1.9 billion records were either lost or stolen, where 18% of breach incidences was due to accidental loss and 8% attributed to a malicious insider. There are challenges to recognizing a data breach that arise from accidental loss, but proper staff privacy and data training can mitigate any potential breaches caused by the fact.

The next step to responding to a data breach incident is to analyse, this encompasses finding the source(s) of the data breach such as vulnerabilities, the intent of the breach or the objective of the attacker, the solution to the data breach or patching the vulnerabilities, and finally lessons learned so it does not occur the same way. Analysing the sources of the data breach can be simplified if the organisation has a proper inventory of the personal data that they hold, and a data flow map, which is covered in chapter 3.

Incident Response Cycle

Responding to the incident is the final stage, and under the GDPR it requires that the organisation notifies the relevant authority so that they can ascertain if any further steps may be required. The incident response life cycle is the execution of an incident response plan, if the plan is the policy that the organisation creates, the life cycle is the carrying out of the plan.

Questions may arise such as why this is considered a life cycle and not a linear model. It may not seem obvious at first, as solutions to problems should be solved in a linear and rational manner, the difference here is that during a security breach it is more valuable to learn from breach rather than simply patching the problem and dusting it off as job well done (hackers can be very dynamic). Therefore, it is important to view the process as a cycle rather than a linear step by step approach.

The National Institute of Standards and Technology (NIST) a branch of the U.S Department of Commerce, operates similarly to the UK based NSCS (discussed in chapter 2), both have developed a cyber security framework that includes an incident response plan which is widely used by many industries globally. The framework holds a certain degree of international standard and is an appropriate model to follow, even for organisations looking to comply with the GDPR. The framework is essentially a fleshed out IRP, this section will dissect each part of the process and its relation to compliance with the regulation.

The NIST framework lays out a 4-step procedure for incidence response the 4 steps are:

1) Preparation
2) Detection and Analysis
3) Containment Eradication and Recovery
4) Post-Incident

To simplify, somewhat ironically, it may be easier to understand if it is viewed as a six-step approach, rather than a 4 step by separating detection and analysis, and containment and recovery into their own bracket (figure 9.2); so, the reworked framework would look something like this:

Figure 9.2: Incident Response Plan NIST

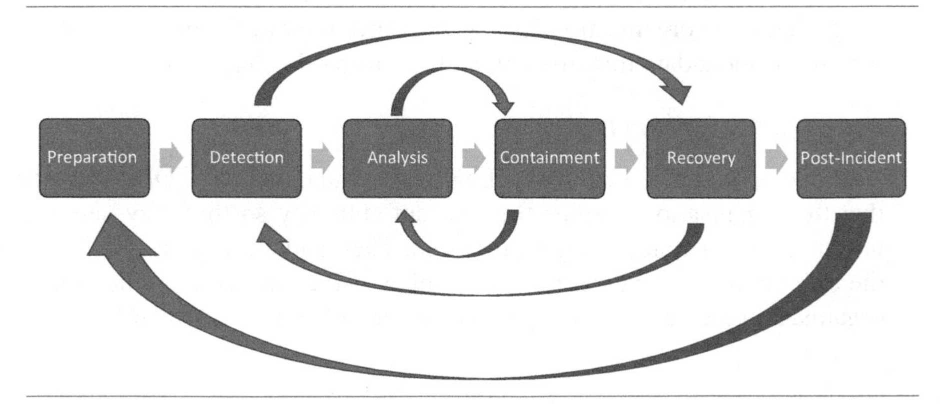

Preparation

Although it is said that detection is the most important phase to incidence response, as it would be helpful to know it is even happening in the first place, preparation is the key phase in how to best avoid and deal with a data breach. Preparation is there to make sure that a) the organisation is ready for the eventuality (as the saying goes it's not if, but when) of a data breach and b) to put measures in place to best avoid a data breach from occurring. This is achieved by ensuring that the networks and vulnerable systems are secured with the best available methods (think: encryption, privacy and protection by design, acute staff awareness, etc.). Although it is important to note that the preparation phase is not typically something the CSIRT or the response team is responsible for, never the less it still is a crucial element in the response life cycle.

The NIST framework outlines resources to better prepare for an incident. This list is adapted to serve in GDPR compliance the first port of contact should always be the data protection officer (DPO, discussed in greater detail in chapter 7). Below are a few examples of the resources necessary to best prepare for an incident under the GDPR.

- Contact information of the organisations DPO and any relevant member the DPO team (if the organisation requires it).
- Ease of communication with any department that is directly or indirectly affected by the data breach.
- Situation Centre: a meeting place for any relevant parties to meet in the event of a breach, the organisation may not have the resources to dedicate a space or the need, but procedures should be put in place to create a temporary space in the event.
- Encryption Methods: Utilizing the most relevant software to encrypt any personal data that the data controller or data processor hold.
- Necessary Hardware: such as;
  - Secure Laptops or work stations
  - Packet Sniffers (analyser)
  - Digital Forensic tools (mainly used during the analysis phase)
  - Spare servers or networking equipment that is not directly connected to the network that has the highest chance of being attacked.

These are just a few of the tools that should be utilised to minimize the exposure to attacks. The key here is to allow for the least amount of breaches, as avoiding them entirely is very unlikely, but minimising them can help the CSIRT or response team better manage a breach situation, especially if they aren't overloaded with breach events.

The two main components in the overall scope of preparation is adequate risk assessment and staff awareness training. These two are generally the hardest to get right but there are many helpful guides and tools that the industry provides to improve in these areas. The other two are network security and cyber attack prevention mechanisms these are relatively easy to implement compared to the other two as they simply require the most relevant tech for the organisation and a system, such as a SIEM (Security Incident Event Management, is a type of software that can assist in detecting security incidences), to monitor network activity.

Detection

Detection can be a little tricky, especially given that the GDPR states that the notification of a personal data breach must be made to the relevant Supervisory Authority within a 72-hour deadline. It is only when the organisation has become aware (or should have known) of a personal data breach does the countdown begin. The Article 29 Working Party defined "awareness" as being:

> *"a reasonable degree of certainty that a security incident has occurred that has led to personal data being compromised"*

This of course is only the case when a Controller or Processor has become aware, with a reasonable degree of certainty.

Detection is not only a difficulty due to tight restrictions placed on the time frame set out by the GDPR, but also a general nuisance for most professionals in the industry. The first step in detection is to have a thorough understanding of the methods of attack, a more in-depth analysis of attack vectors was covered in chapter 4. In brief some examples of attack vectors that an organisation may experience are;

- Malware attacks: Ransomware etc.
- Phishing attacks: Impersonation, Vishing, etc.
- External/Removeable media: A form of phishing that involves leaving USB sticks around the office space in the hopes that some one will plug it in.
- Email: malware or phishing carried out over email.
- Dumpster Diving: Looking for critical information about the organisation by scanning waste paper deposits or the like.

There is a plethora of options that an attacker can utilise to gain unauthorised access to an organisations system which cannot all be named here, but it should be noted that the organisation should be flexible enough in their attitudes toward security to adapt to new threats and to facilitate ease in the detection of intrusion.

The signs of an incident can come one of two ways, either as a precursor or as an indicator. The precursor is a signal that an incident may happen in the future, while an indicator is a sign that an incident is occurring currently or has already occurred. There are a few reasons why detecting an intrusion can be difficult for an organisation a couple mentioned are;

- High number of possible attack vectors
- The skill and focus of the attacker (a ratio combination of high and low)[1]
- Organisations may not have the deep technical knowledge to effectively analyse the network.
- Traditional intrusion detection software may depict high volumes of alerts but may be showing false positives[2]

If any signs of a precursor are present the organisation has an opportunity to stay ahead of the attack and may even prevent it from occurring. Although these instances are rare, the elements an organisation should be looking out for are:

- User log entries: anything suspicious, this could be employee login attempts who no longer work for the business (expired accounts for example).
- Bandwidth usage: there may be spikes at times of attacks (especially during DDoS attacks), it is particularly important to analyse weather the usage differs from average use on any given day. Example, on Sundays the organisation is not open, but bandwidth usage is unusually high.

---

[1]As an example, an attacker of low skill and low focus may be using scripts to attack a large array of networks, if the relative security of the organisation is better than their neighbours there is low chance the network will be harmed. On the other side a high skill high focused attacker may personal target the organisations network regardless of the security measures in place, usually using APT's (Advanced Persistent Threats).

[2]This could be related to the way employees interact with the network, if not adequately trained may misuse a certain tool resulting in the intrusion software flagging the user.

- Any threats from criminal groups or hacktivist that are targeting the organisation. Although some may be empty threats these are still important precursors to acknowledge.
- Any cyber-attack prevention software such as, antivirus, intrusion detection software, firewalls etc. alert the administrator.
- Failed login attempts, the network system may notify the admin with multiple failed login attempts into the organisations networks.

These are a few examples of what is considered to be a precursor to a potential security incident. Indicators work similarly to precursors, but the organisation will have less time to react as they are used to detect if a network is currently being attacked. Indicators will mostly be found by antivirus or the SIEM software that will alert the systems administrator of a potential incident.

Analysis

Analysis in the incident response plan is the weeding of detection's garden. If the detection phase is the cultivating of a garden, analysis is making sure the gardener removes the weeds and not the crops. Although this metaphor makes it seem simple when it is very easy to tell difference between a crop and an unwanted weed, in the cyber space is it not so simple.

There are many variables to attend to when accepting a detection event as a real incident. Detection events, that are flagged by some systems, may often appear at first as a credible threat but, may be a false positive. An example of this may be a user forgetting their password an entering it multiple times, the system could potentially flag this as an incident. This poses a challenge to security professionals and GDPR practitioners as it could involve investing a disproportionate amount of time responding to false threats rather than dealing with actual threats.

This is where refining this phase, and why it feeds back into the containment phase, is crucial to the organisation (after containment it is useful to analyse if any other connected systems may be affected by the same issue and rectifying where necessary). The security discourse should be an ongoing conversation and that ethos should also apply to how the organisation responds to incidents.

In the detection phase there were two main signs of a potential event indicators and precursors. The goal of the analysis phase is to ascertain the accuracy of the indicators and precursors. If the accuracy could be guaranteed to be genuine the analysis and human involvement would be unnecessary.

The trouble is that even if an indicator is accurate, there could be multiple reasons for that, but not all of them may be a security incident. For example, there is a system crash due to an intern accidentally changing a configuration file the incident would be flagged but it would because of human error and not a malicious attack. This is where the security team would analyse the event and act accordingly.

There is a simple strategy for analysis of a situation known as ODDA loop. The OODA loop (pronounced oh-da loop) was first developed by United States Air Force Colonel John Boyd (seen in figure 9.3). The strategy centred around dogfights, or combat engagements in the sky, the process involves observing, orientating, deciding, and then acting. This process has been adapted to the business world and health care professionals too, and in turn the cyber security space. In the analysis phase when events are flagged it is beneficial to employ the OODA loop process to ascertain the accuracy.

Figure 9.3: The OODA Loop

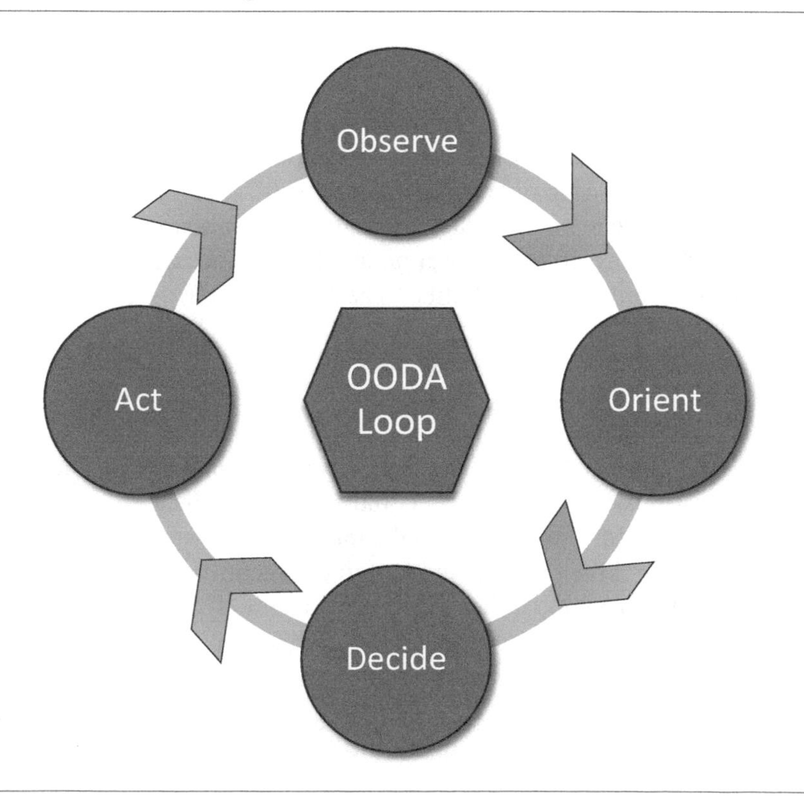

This process can help identify events that are credible, it can also help detect outliers where no action is needed.

- *Observe:* This part of the process is a version of the detection phase of the incident response plan. The first stage is to notice if a problem has even occurred meaning the security team must remain vigilant and respond accordingly.
- *Orient:* Is part of the analysis of an event, this part of the process should be used to figure out the accuracy (i.e. is it a security incident or not) of the event and the cause of the incident. Orientate towards the location of the event and any possible explanation of the event.
- *Decide:* Here the team should decide weather the event warrants further investigation, and asses the risk of not dealing with the event.
- *Act:* This is when the team should act upon the decision that was made about the event. Example, the team has found spike in network traffic and have decided it was a DDoS attack then in this phase they should act appropriately according to the decision made.

The OODA loop can be used as a micro-decision-making process in each phase of the incident response plan.

There are some recommendations and tools that can assist in making analysis simpler. There are a few obvious indicators that an event has occurred and those involve their impact to the CIA[3], discussed in chapter six. The response team can analyse weather any or all the aspects of the CIA were affected to ascertain weather an incident has occurred (example has the breach affected the confidentiality? The Integrity? Or the Availability of the data?).

The other not so obvious ways but helpful ways to find the effects of an event is to:

- Know normal behaviours and patterns of both end-users and staff. Deviations from the norms could indicate an incident is occurring or has occurred. This also is applicable to the network or systems regular behaviour which is more evident to the security teams than the psychological behaviour of the organisations staff. There is no full proof way

[3] As a reminder the CIA is the confidentiality of data, the integrity of data, and the availability of data (the organisation can think of this as a product of service too, ex. Availability of the service they provide, or the privacy afforded to the users).

of knowing all behaviour, which is why the team should be dynamic in their structure filling knowledge gaps with relevant experts.

- Have a system which can profile the network or information system. These are characteristics of the system expected activity so that changes can be easily identified. This also ties into the CIA of the system, having a checklist that outlines those elements of the system so that checking for deviations from them might indicate an incident.
- Create a policy that collates all events and logs them in one place. This can facilitate in analysis as any department can report to one place and documentation is done in a single format.
- Use packet sniffers and other detection tools to generate a higher volume of detailed information of a potential data breach or incident event. Change around the settings of the software so that it is more adapted to the organisations network patterns and activities or information gathering tools.
- Synchronize host clocks, analysing events and trends with past events could pose a challenge if timing is inconsistent in the network. In the previous chapter this is described in a little more detail as it is one of the international standards in the ISO, ISO 8601 which describes accepted standards for the format of time keeping.
- Share information, a significant reason attackers are generally highly successful in penetrating systems is because they tend to share information amongst one another. This creates a space of heightened awareness amongst the hacking community lending knowledge to a wide amount of systems, that other organisations might not know. This is not only a key to success for attackers but for organisation too, a helpful tool in analysis is to seek help and interact with the security community at large. This notion can aid at stopping future incidences or from other organisations experiencing similar issues.

It is the analysis phase that the organisation should keep a well-defined documentation of any events that were further investigated. A simple log book would do the job, but the organisation should have a policy in-place to ensure that all departments adhere to an accepted format. The GDPR does not outline a format or process for documentation in the event of a personal data breach, but the organisation should have methods to ensure that compliance to the regulation was met and all appropriate measures were taken during the incident. The things that should be included are cited in article 33(3):

The notification referred to in paragraph 1 shall at least:

(a) Describe the nature of the personal data breach including where possible, the categories and approximate number of data subjects concerned, and the categories and approximate number of personal data records concerned;

(b) Communicate the name and contact details of the data protection officer or other contact point where more information can be obtained;

(c) Describe the likely consequences of the personal data breach;

(d) Describe the measures taken or proposed to be taken by the controller to address the personal data breach, including where appropriate, measures to mitigate its possible adverse effects.

Article 33(3)

After analysis is completed and the event is found to be a security incident, and the rights and freedoms of the data subjects are in jeopardy it is at this point the organisation has become aware of the breach and should respond accordingly. As mentioned at the beginning of the chapter the organisation must notify the Supervisory Authority within a 72-hour period.

Containment

After a breach has been found and the organisation has notified the relevant authority measures should be taken to contain and mitigate the effects of the breach. A containment strategy should be employed before an attack completely overwhelms the system. The strategy will depend on the type of attack the organisation is suffering, an attack deriving from an SQL injection will require a different approach compared to a loss of availability due to a network DDoS attack as an example.

Considering the unpredictability and scope of attacks the organisation should have a strategy that fulfils certain criteria to determine the correct strategy:

- Availability to the service.
- Protecting confidentiality
- Protecting Integrity
- Investment of time and resources to implement the strategy.

A technique of containment that is often used by Security Operation Centres (SOC Labs) is sandboxing. Sandboxing allows security teams to release a malicious code in a vacuum, essentially the team can contain and execute malicious code in a safe environment. For example, malware can be tested and activated in a sandbox to analyse the possible effects on the system without it interrupting or infecting other systems.

There can be issues with sandboxing and containment methods that involve disconnecting the infected systems from the network. The disconnect may result in the malware encrypting the host hard drive data rendering it unavailable to the user or organisation, causing more damage when contained. Therefore, it is important to note that merely disconnecting a system may not be sufficient enough to contain the incident.

Eradication and recovery

After an incident has been contained the eradication process begins. In this phase the security team should comb through the rest of the networks systems and delete any unwanted files and possible malware packages. This could also involve disabling affected user accounts, identifying vulnerabilities in the systems, and patching them accordingly.

Finally, the team should then restore the systems back to its normal functions, resetting database where necessary. This may be done by restoring the system from a backup file or to a previous state (i.e. before the malware was introduced), it also involves ensuring the systems is functioning normally. The recovery may not be as simple as restoring the system from a backup it may be far too damaged resulting in having to rebuild the system from the ground up.

In the recovery phase it is important to make sure all security protocols are in place, start with the basics such as firewalls, encryption, access controls, etc.

Post-incident

In this phase of the incident response plan the security team and the organisation should review the incident and learn lessons from it. It is crucial that the departments of the organisation work together to fully document the event, outline the vulnerabilities and points of access, and ensure the same attack cannot happen again.

Questions that should be answered in the post-incident meeting and assessment:

- What happened for the breach to occur.
- Evaluate the performance of staff and management of handling the incident.
- What can be done to prevent this from happening in the future.
- What can be done differently if the event were to happen again.
- What Indicators and Precursors should be analysed closer.
- Are there any tools or preventative software that the organisation should invest in.

The post-incident has other benefits such as giving the organisation material for training their staff, or opportunities to update organisational policies and procedures ensuring a mature attitude toward security.

Notification for a Personal Data Breach

The GDPR simplifies the notification process as Data Controllers are only required to notify the relevant Supervisory Authority in one country. This is generally taken to be the EU country in which the organisation has its registered office. In the UK this is the Information Commission Office. Article 33 of the GDPR outlines the requirements for notification; essentially this involves notifying the authority within 72 hours of becoming aware of the breach (Article 55), with a notice documenting the above-mentioned elements. The data controller is required to document all breaches of personal data; the document must contain the facts relating to the breach, its effects, and any remedial action taken to mitigate the consequences.

The Supervisory Authority may give guidance and advice to the controller on how to proceed, including compelling him to notify the data subjects. Where the breach is notified to the Authority outside the 72 hours of discovery, it must be accompanied by an explanation for the delay. This may or may not be accepted by the authority who will then decide on any punitive measures to be applied to the organisation. Article 33 (4) allows for the notification to be delivered to the authority in phases, where it is not possible to provide full information all at once. As in the case of an ongoing breach.

Where the Data Controller has applied technical and organisational controls to the data which has thus rendered it unusable to any third party, and the risk to the data subject's rights and freedoms have been mitigated, then only notification to the supervisory authority, and not the data subjects, is required. The requirement to notify the data subjects is at the discretion of the supervisory authority, regardless if the above conditions are met. Where a notification to the data subjects is deemed necessary, the controller is obliged to make the notice clear. It must be written in plain language stating the nature of the data breached, and the information and recommendations based on Article 33(3): mentioned above.

Data Breach Severity

Assessing the severity of a data breach can be an arduous but necessary process after having responded to an incident. The benefits of assessing the severity are primarily for the Supervisory Authority in ascertaining weather appropriate measures were taken by the organisation to best counteract the security incident, and for the organisation to review the potential financial damage that occurred as a result of the breach. It is also beneficial to the organisation in assessing the extent and the recovery time of the systems affected by the breach.

Within the context of the GDPR, the risk posed to individuals (the Data Subjects) and their Rights and Freedoms, by a personal data breach is defined as the Likelihood (probability) and the Severity (degree of damage) of any negative consequences. These two elements are key in assessing the extent of the breach.

Assessing the severity and likelihood of a security incident involving a breach of personal data is fundamental to understanding the consequent relevant responses required by the Data Controller, the organisation, and any third-party data processors involved. Unfortunately, the GDPR does not state the way in which this assessment should be achieved, only that activities which could be considered "high risk" must be properly managed. The GDPR highlights the number of data subjects involved (volume) and the potential harm to them through loss of confidentiality, especially where discrimination, social, or economic disadvantage may result, as being critically important.

There is no currently accepted industry-wide data breach assessment protocol, however taking note of those industry sectors which process "high risk" personal data and analysing suitability (as an industry use-case application) may provide a solid starting point. Industries which engage in the provision of health care, services to vulnerable groups (children etc), criminal justice services, National Security, certain financial services, and large scale public monitoring (such as traffic cameras) may automatically come under the heading of "high risk".

Article 33 of the GDPR provides three examples of high risk activities:

1) automated profiling that is systematic and extensive and which significantly affects individuals,
2) large-scale processing of special categories of data, and
3) large-scale, systematic monitoring of a publicly accessible area.

Clearly, those activities involving large-scale processing of personal data and automated evaluation of the personal characteristics of data subjects are more likely to be considered high risk and thus impacting the severity of a breach. The GDPR emphasises the high level of risk where large numbers of data subjects are included in a breach.

Article 33(3) of the GDPR outlines some of the requirements to be considered when determining the risk posed by the breach. Specifically, it asks that the nature of the personal data breach including the categories and number of data subjects involved and the categories and number of personal data records concerned is detailed. It also asks that a projection of the likely consequences of the breach be provided.

When attempting to assess the impact of a personal data breach it is important to use a systematic approach. Questions must be asked on a case specific basis:

How many data subjects are affected by the breach? Important in assessing OVERALL risk. The more data subjects affected the greater risk. Many breaches involve hundreds of thousands of data subjects, for example the Equifax data breach of 2018 affected around 11 million users which was considered extremely high in severity due to the number of data subjects involved.

Breach Severity Rating and Risk

Likelihood

What is the probability that the breach will negatively affect the data subjects? The primary elements that should be considered under the GDPR are:

What has happened to the data?

- Damaged: If the data was damaged physically, as in the case of a power surge causing hardware damage, the probability of recovery is low, and the highest risk would be attributed to the availability.
- Lost: Inadvertent loss of data through system failure or physical "disappearance" as in cases of USB's falling down the back of the sofa, the specific circumstance under which the data was lost would vary the likelihood of risk to the rights and freedoms of the data subjects.
- Stolen: Malicious appropriation of the data has a high likelihood of negative consequences to the data subjects and would be considered high severity.

Regardless of what has happened to the data it is important to assess what, if anything, a third party may extract from it, and how might it be used.

Clearly the role of encryption, and other protections is important here. If the data is otherwise unusable the probability of certain types of damage to the data subject is reduced. However damaged data may infringe the data subject's Right to Access and control of their data as illustrated in the case of the damaged data, although this can be mitigated by keeping backups. As the assessment of the severity of the breach needs to be done on a case-by-case basis, and time is limited, it is vital that an assessment tool or protocol be used.

A useful way to assess severity, discussed briefly in the previous sections of the chapter, is to analyse the affects on the CIA confidentiality, integrity, and availability of data which play a large role in the ENISA methodology of data breach severity calculation.

Confidentiality

Has the security incident affected the confidentiality of the data? Article 33 of the GDPR states that when a breach of confidentiality is detected it must *"without undue delay"* be reported to the supervisory authority,

"...unless the data breach is unlikely to result in a risk to the rights and freedoms of a natural person".

To assess the severity on confidentiality, it is useful to know what type of data was affected by the security incident. There are two main types of data sensitivity that the organisation should be concerned with when attempting to assess the severity: very sensitive data (such as medical records), and what would happen if the data were to be misused due to a breach in confidentiality (e.g. for financial gain).

The next element to assessing data breach sensitivity, which applies to all the CIA, is the data subjects involved and their types (i.e. special categories):

- Vulnerable Groups (special categories of data)
- Employees
- Customers
- Service users
- Suppliers
- Etc.

Integrity

Any unauthorised changes to the data may lead to damage to the DS through consequences such as Identity theft or Fraud, damage to reputation or standing, or wider Social and Economic damage. The breach severity, when dealing with data integrity, can be assessed by analysing the circumstances in which the breach took place.

Things to consider are:

- Were there protection in places; like encryption, pseudonymisation (or full anonymisation), staff awareness, etc.
- What were the causes of the data breach: was it lost, damaged, or was it maliciously acquired?
- What could the data reveal to a third party: how can it be misused if changed?

The details to consider for the integrity of data is, has the breach resulted in any unintended changes to the data (data corruption)? This is not only limited to the accuracy of data but it is also related to the retrieval and storage of data.

If the breach has resulted in difficulty in any of the elements (retrieval, storage, or accuracy) of data it is likely to score high on the breach severity assessment.

Availability

The last part the organisations should be looking at to evaluate the severity of the breach is to asses the impact the breach had on the availability of data. The CIA has been discussed throughout the book but in the context of breach severity the impact on availability of data will greatly affect the severity of the breach. For example a loss to availability of data for an organisation dealing with sensitive health information could cause serious physical damage if the data was unavailable for extended periods of time, compared to a company that might experience web traffic loss to its website if they are suffering a DDoS attack.

The two examples given would result in very different breach severity scores as one is dealing directly with the physical well being of the data subject, whilst the other might only be affecting data subjects convenience of visiting the website (although considering the regulation this case might not affect the rights and freedoms of the data subject and would not register as a data breach worth notifying to the Supervisory Authority).

Testing for availability is as simple as making sure the organisations network is online and has not been disrupted by either accidental or malicious means. The Security Information Events Management, which is a software employed by many cyber security professionals, can help not only in detecting potential data breaches but is also a way of keeping track that all the systems are operating as normal, meaning it is possible to test the availability by monitoring the SIEM.

ENISA Methodology

ENISA is the European Union Agency for Network and Information security, they were set up in 2004, with their headquarters located in Greece. The agency works with the member states and private institutions to give advice surrounding cyber security. They have establish a widely used method for calculating the severity of a data breach, broken down into three components.

1) Data Processing Context (DPC): here the type of data being processes and the nature of processing are identified and weighed. An example of data type and score would be;
   a. Data type=Financial, Score=3 the higher score indicates more sensitive data, simple data sets such as names would not score as

high as the worth to malicious actors is less than having financial or medical data for example. The score can change even on simple data sets if they contain elements that could enable assumptions of the data subjects financial status or health condition pushing the score up, the opposite can happen if data does not contain insight into the data subject behaviour. The maximum score a DPC can attain is four with the minimum being 1 (anything lower i.e. 0.5 must be justified). *See appendix for more details.*

2) Ease of Identifying the data subject (EI): Ease of identity is fairly straight forward the score is dependent on how easy it is to identify a data subject from the breached information. The score is widely dependant on how much the organisation has done to 1. Protect the data through pseudonymisation for instance and 2. The amount of systems affect by the breach that could lead to indirect identification from inference.
   a. The score is on a scale from 0 to 1, 0-0.25 being negligible meaning it is very difficult to infer or match the data subject to a particular person but may be under very specific circumstances. To 1 being the maximum were very little research is required to figure out the identify of the data subject.
3) Circumstance of Breach (CB): Finally the last component is circumstance of breach. This component asses the actual occurrence of the breach and the situations leading to the breach. This surrounds the CIA of data covered in the above section. The score is altered depending on the reason the breach affected one of the elements of the CIA, for example was it malicious or accidental loss, destruction, etc. The scores are generally weighed between 0.25 to .5 depending on weather it was malicious or the degree to which it has affected CIA (example Availability is only temporary = .25 compared to Availability is greatly affected meaning data is unrecoverable = .5).

Finally after figuring out the scores of each individual component the organisation can calculate the overall severity rating of the breach given by the formula below:

$$\textbf{Severity Rating} = (\textbf{DPC} * \textbf{EI}) + \textbf{CB}$$

The severity rating will be given on a 4 point scale with a severity score of 2 or less equalling a low rating, to 4 showing a very high severity score meaning the breach resulted in very high risks to the rights and freedoms of the data subjects. A very high severity rating could also mean irreversible consequences for the organisation financially, reputationally, etc. which could be extremely difficult to overcome.

Severity Score	Rating	Description
SE < 2	Low	Individuals either will not be affected or may encounter a few inconveniences, which they will overcome without any problem
2 ≤ SE < 3	Medium	Individuals may encounter significant inconveniences, which they will be able to overcome despite a few difficulties.
3 ≤ SE < 4	High	Individuals may encounter significant consequences, which they should be able to overcome albeit with serious difficulties.
4 ≤ SE	Very High	Individuals may encounter significant or even irreversible consequences, which they may not overcome.

Chapter Review

Summary of Key Points

1) Components of CSIRT were covered, along with the difference of an internal and external CSIRT.
2) The elements and procedure of an IRP and the incident life cycle.
3) Data Breach Severity: the effects on the CIA and how to calculate using the ENISA methodology.

- Think about how best to set up a CSIRT team which departments would you include. How would you decide?
- Consider the application of an incident response plan beyond a cyber security event.
- Would you include anything else in a data breach severity calculation?

References

Caralli, R., Allen, J., Curtis, P., White, D., & Young, L. (2010). CERT ® Resilience Management Model. *Management.*

Cichonski, P., Millar, T., Grance, T., & Scarfone, K. (2012). *Computer Security Incident Handling Guide: Recomendations of the National Institute of Standards and Technology.* NIST, Gaithersburg.

D'Acquisto, G., Domingo-Ferrer, J., Kikiras, P., Torra, V., Montjoye, Y.-A., & Bourka, A. (2015). Privacy by design in big data: An overview of privacy enhancing technologies in the era of big data analytics. *Enisa.*

ENISA. (2013). *Recommendations for a methodology of the assessment of severity of personal data breaches.* ENISA, Heraklion.

Horne, B. (2014). On Computer Security Incident Response Teams. *IEEE Security & Privacy.*

Mandia, K., Prosise, C., & Pepe, M. (2003). *Incident Response & Computer Forensics.*

Massey, S. (2017). *The Ultimate GDPR Practitioner Guide.* London: Fox Red Risk Publishing.

Mepham, K., Louvieris, P., Ghinea, G., & Clewley, N. (2014). Dynamic cyber-incident response. *International Conference on Cyber Conflict, CYCON.*

Mitropoulos, S., Patsos, D., & Douligeris, C. (2006). On Incident Handling and Response: A state-of-the-art approach. *Computers and Security.*

Wheatley, S., Maillart, T., & Sornette, D. (2016). The extreme risk of personal data breaches and the erosion of privacy. *European Physical Journal B.*

Appendix

Appendix 9a: Example of DPC Component of the ENISA Methodology

Data Type	Description of Processing	Score
Simple Data sets	Professional, Name, Geolocation data	1
Behavioural	Location, Traffic data, data on personal Preference	2
Financial	Any kind such as credit information, income statements, social welfare	3
Sensitive Data Sets	Health, Personally identifiable information, Social security numbers	4

Appendix 9b: Example of EI Component of the ENISA Methodology

Level	Description	Score
Negligible	Extremely difficult to match data to a particular person	0.25
Limited	Matching data to a person is possible with additional data sources	0.50
Significant	Identification is possible indirectly, the breached data plus some research would be sufficient	0.75
Maximum	Identification is Possible directly from the breached data	1.00

Chapter 10: Valuing Security

At a Glance:

- Making the business case for security
- Budgeting methods and use case for IT
  - Run, Grow, Transform (RGT)
  - Mapping the Budget
- Annual loss expectancy, Return on Investment, Asset value calculations for cyber security
- Effective Communication
  - Talking to Management
  - Presentation Structure

Learning Objectives: *Students should be able to…*

- Explain the use and value of effective cyber security for the organisation.
- Understand the budgeting process and how it relates to IT and security.
- Give examples of the do's and don'ts of presentations.
- Evaluate the cost of cyber security and compare it to the value of implementing it.
- Understand some basic financial principles related to communication with management.

Key Terms

1) Budget Assessment
2) Annualised Loss Expectancy
3) Asset Value
4) Return on Investment
5) Prospect Theory
6) Market for Lemons
7) 10/20/30 Rule

Valuing Security: Making the Business Case

Cyber security is not simply an IT issue, it is the security for the whole infrastructure of an organisation. As such the budget for cyber security should be drawn from every departments across the organisation; sales, personnel, IT, admin, accounting, governance, marketing, senior management, etcetera. In reality, it is most often bundled into the IT budget and so it becomes the responsibility of technology experts rather than business or governance specialists. The issue which arises, from the mis-placed idea that cyber security is *just* an IT competence, is one of accountability and transparency. The GDPR makes it very clear that senior management are accountable for failures in security that impact the personally identifiable information of data subjects, and it is important that IT professionals understand this and communicate it to the rest of the organisation.

A good place to start is with the finance department, since they control the flow of funds through the organisation. The Chief Financial Officer (CFO) has a key role in the allocation of departmental budgets and it is the CFO who often approves or rejects requests for the funding of projects. This being the case, it is vital that IT professionals develop an ongoing relationship with the CFO, based on regular, frank, and informative communication. The language of this communication must be one the CFO understands; the language of business, and the language of business is finance.

The CFO is an expert in finance and accounting and is responsible for financial planning, which is used to mitigate the exposure of the organisation to financial risks. As a senior manager the CFO will be involved in high-level strategic decision making and understand the financial health of every part of the organisation. Knowing where the money is flowing to, and from, gives the CFO insight about the organisation's overall budgetary needs, however it does not give detailed information on specific or future requirements.

The Chief Information Officer (CIO) is an expert in information systems and technology and is responsible for the smooth and secure operation of the organisation's IT, usually including cyber security (C-Suite). It is the CIO's responsibility to help the CFO understand the impact, financial and otherwise, of under-investment in the organisation's IT infrastructure. This impact should be illustrated by explaining the evolving nature of the threats faced, and how these threats would impact each of the major business processes. The potential impact on reputation, productivity, share price, litigation, and regulatory compliance, has a financial cost associated to each, and it is against this potential cost that the CFO will calculate the budget for cyber security.

What is the value of cyber security in an organisation? This is an important question to ponder before constructing a budgetary framework as there are many factors to consider, only one of which is cost of technical implementation. Cyber security protects the infrastructure upon which the organisation depends; not just the data it uses. Data has a value, which may be quantified in monetary terms, but it does not reflect the value of the infrastructure as a whole. Cyber security is, on the one hand, a type of insurance, and on the other, an integral part of the revenue generating process. In the insurance function it fulfils risk management and loss prevention requirements; protecting and preserving the integrity of the organisation's most important assets, the loss of which could cripple its ability to function.

The primary role of cyber security in the revenue generating process is as an ancillary support to the sales and marketing departments. Security issues which lead to impaired functionality in these areas have a direct effect on revenue, usually through the impact of system downtime on sales and reputational damage to the organisation.

Budgeting for IT and C-Suite

The IT budget is primarily a tool with which CIO's can show how the organisation's IT function contributes to the business value. The budget is a management tool which gives senior management insight into the contribution IT makes to the organisation's ability to achieve the desired outcomes; in a language they can understand. It is also an accounting tool, which estimates and controls how much money will be spent, and how and when that expenditure will happen. Done correctly, an IT budget should show how the contribution by IT and cyber security, toward the achievement of the organisation's strategic goals, is delivered. This approach presents IT and C-Suite as helpful resources within the organisation, rather than as a pure cost centre, and the focus on the achievement of strategic goals, rather than technical needs, will increase the likelihood of receiving budget approval.

CIO's are not expected to be experts in accounting and finance, but they are required to formulate their departmental budget. When faced with a new area of expertise with which one is unfamiliar, it is prudent to speak to someone who is an expert first; in this case an accountant. Developing a relationship with the CFO and with the accounting department is crucial for gaining budgetary approval; as the CFO is the person responsible for signing-off on organisational expenditures. An IT department that understands how and why

Figure 10.1: IT budget assessment

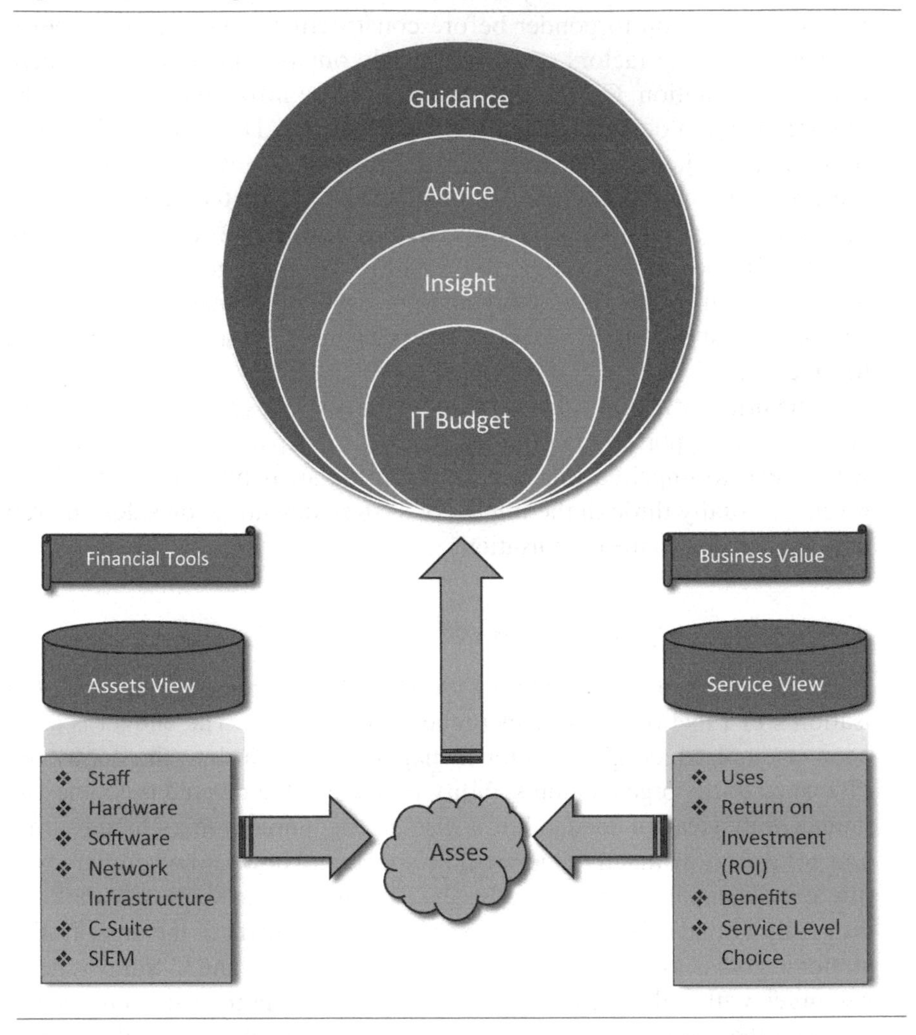

the CFO makes decisions on budget requests, is a better funded IT department. For example, ask the accounting department for (see figure 10.1):

- *Guidance* on budget preparation; especially regarding the preferred accounting techniques, which have specific methods for the calculation and categorisation of different types of expenditure.

- *Advice* on payment options for large infrastructure expenditure or high-cost projects. A request for costly server upgrades is more likely to be approved if the cost is spread out over a number of years.
- *Insight* about the planned timing of major expenditures within the organisation. Knowing that a major building project is planned to begin in 6 months' time may inform the timing of IT budget requests; usually before other major projects.

Understanding what a budget is, and how it should be constructed, is useful; regardless of the context in which it is applied. It provides detailed and clear information on which to base spending decisions. A broad outline of the budgeting process and some of the relevant finance tools, is followed by an introduction to presentation skills. The budget can't sell itself; it needs to be explained and justified in a way the audience (senior management) can engage with and support.

Budgeting

> *A quantitative expression of a plan for a defined period of time. It may include planned sales volumes and revenues, resource quantities, costs and expenses, assets, liabilities and cash flows.*
>
> *– Certified Institute of Management Accountants, 2005*

A budget is a financial plan which estimates the revenue or expenditure of a project or an organisation, over a given time period. Budgets delineate what is possible and desirable for the effective achievement of an organisation's objectives as they relate to income and expenses. Most organisations will draw up a budget at least once a year and this will be for the organisation as a whole, and for the separate departments, divisions, and subsidiaries. Departments responsible for income, such as the Sales department, will often be the first to draw up a budget to provide the financial benchmark from which expense budgets, such as personnel, may be set. An organisation must know its spending limit before it can allocate funds between various departments.

Development Process

Since a budget is a forecast it is based on a set of assumptions about the future. These assumptions are based on previous periods of economic performance, both of the organisation and the environment in which it operates. The trends in sales and expenditures are projected into the forecast period, and financial estimates are derived. The departmental budgets are published with an explanation of the assumptions upon which the figures are based. The individual budgets are then incorporated to form the master budget; the organisational budget, which is then reviewed by senior management before receiving approval.

There are two types of budget; the fixed or static budget, and the more flexible continuous budget.

Fixed budgets: are set at the beginning of the forecast period and do not change, regardless of the actual volume of income or expenses.

Continuous budgets: allow for regular updates in actual income and expenses to be incorporated into the budget and is often used when accurate forecasts cannot be made. Within an IT context, the Run, Grow, Transform, (RGT) approach to budgeting provides an opportunity to breakdown the budgetary requirements into these three categories, as seen in figure 10.2.

1) *Run*: these are mission-critical aspects of IT that allow the organisation to function at its most fundamental level. This includes all elements, hardware, software, and personnel, without which the organisation would stall. This part of the budget should not be reduced or cut under any circumstances. Run items are part of the organisational infrastructure.
2) *Grow*: these items bring additional advantages to the organisation, through increasing efficiency, improving security, and providing capabilities through which the organisation can grow. These items can be termed as cost-reducing or revenue enhancing; such as an

Figure 10.2: Run, grow, transform

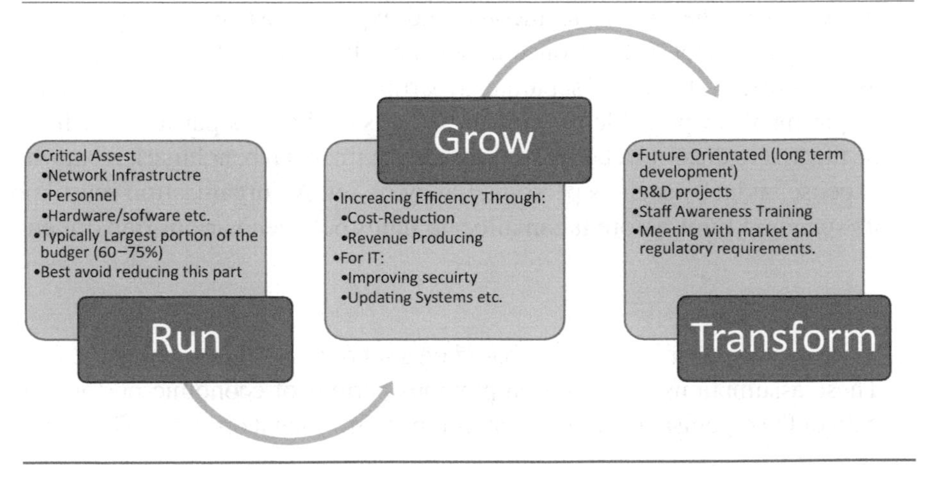

upgraded website which increases customer spending, or a firewall which reduces downtime.

3) *Transform*: these items are future oriented and deal with where the organisation is heading; its long-term development and sustainability. Essentially involving research and development initiatives which investigate upcoming technologies and organisational needs. These items may also include education and training.

Breaking the budget into these categories allows for the assessment and analysis of spending levels as a percentage of the total budget. The Run category is always the largest, taking between 60 to 75% of the total budget. However, as technology evolves to become increasingly efficient this percentage reduces over time, leaving more room for the Grow and Transform categories allowing for the provision of future growth and success. A Run, Grow, Transform, analysis looks at the IT budget relative to the goals and objectives outlined for the department, and is able to highlight areas of weakness, or overspending.

A structured approach is best when preparing a budget, there are four key steps: define the goals, review the existing situation, define the costs, and draw up the budget.

Define the Goals: Use the RGT method to map the organisation's goals and objectives as they apply to IT. Analyse the organisation's strategic plan and communicate with relevant stakeholders to gain insight into what is needed. Once these needs are identified, the organisation should conduct a comparative assessment of the performance of the IT department and the organisation against that of the industry. This will provide a benchmark for the level of spending requested, matched to the RGT categories, which the senior management can readily understand.

Review the existing situation: Once the goals are defined it is necessary to assess the degree to which they can be met by the existing IT infrastructure. A Gap analysis is useful for identifying areas for investment and expenditure and will provide a starting point for determining the costs involved. Budget drivers are those factors which are specific to the type of organisation and the environment it operates within. In the context of cyber security some of the factors to consider are:

- *Regulatory requirements*: Different industries may have to comply to specific regulations. Compliance and oversight may require a separate line entry within the budget or may be contained within a different budget altogether.

- *Market position*: Where organisations are seen as high value targets they may require greater security measures to offset their large attack surface. A strong cyber security posture can be part of a reputation building strategy which attracts new revenue streams.
- *Sector*: The type of industry informs the ways data is used, processed, and stored. Thus, each industry will have certain aspects of its operation which are unique or idiosyncratic to it. These must be understood before budgetary decisions are made, to avoid unnecessary or useless expenditures. Some industries suffer from attacks on assets in Intellectual Property, and these require more sophisticated and stringent security protection.
- *Audits*: Any findings from internal and especially external audits, that are outstanding must be rectified as a priority. This is presented as a non-negotiable line item in the budget.

Define the costs: Prepare a table of all the items required to fulfil the needs, identified in the first step, which cannot be met with the current situation; those identified through the Gap analysis. It is good practise to seek quotes from at least two vendors for each item, based on like-for-like products and services; where the organisation has a preferred vendor list it is still advisable to attain an outside quote for comparison. Enter these quotes into the table to show management where and why costs may vary; especially where there is a significant difference in price.

Draw up the budget: Once all the required information has been collected and analysed the budget can be created. This process is detailed below.

Mapping Out the Budget

Step 1: What security capabilities are required?

When planning the budget all security capabilities must be included: prevention, detection, containment, response, and eradication; these are discussed in detail in chapter 9. Some of the items to be considered are;

- *Prevention*: training and staff awareness, firewalls, data loss prevention, anti-malware, encryption, anti-spyware, intrusion prevention.
- *Detection*: threat monitoring and detection, insurance for prevention, managed services, intrusion detection system (IDS),
- *Containment*: security architecture, network segmentation.
- *Response and eradication*: breach notification costs, cyber-insurance, personnel, resources for clean-up.

Having considered the above, a risk assessment should be carried out to ascertain the areas of greatest risk. The services of an external third-party subject matter expert will provide insight into areas of risk that have been missed and gives added authority to the decisions contained within the budget. The areas of greatest likelihood and severity are first to be allocated within the budget.

Step 2: Which costs to consider?

The first thing that comes to mind when considering the costs of cyber security may be technological in nature, however there are other factors which are just as impactful.

- *Operations and maintenance:* Staff to manage and maintain the security stack, full-time and part-time staffing for 24 hour a day, 7 days a week monitoring, third-party security provision, off-site storage monitoring, ongoing staff training. It is important to note that third-party services become an operational expense rather than a direct cost, that is an ongoing expense used to maintain the functionality of the organisation.
- *Governance:* The development, creation, and implementation of policies incurs costs through; allocation of staff time to policy making, hiring experts in policy design, implementation requirements such as training and communication. Policies and procedures must also be monitored and reviewed regularly, and this also incurs similar costs.
- *Hardware: S*torage infrastructure and secure rooms can be a significant cost. Also consider security appliances, Storage area networks (SAN), mobile devices, upgrades and replacements, back-up hardware, data pipeline, end-of-life recycling or destruction.
- *Software:* Ongoing license fees for existing and new staff, Security Information and Event Management (SIEM), cloud services, Software as a Service (SaaS), new software.
- *Expect the unexpected:* Reserve up to 20% of the budget for the things that could, and will, go wrong; also called the contingency plan.

Step 3: Categorise the cost

Categorise the cost items into the Run, Grow, Transform, groupings to clarify which items are mission-critical and need to be prioritised above the rest. Every item must be placed into one of the RGT groupings so that its role in the organisation's ability to achieve its objectives is clear.

Step 4: Determine the risk appetite level

That is the level of risk acceptable, as opposed to the level of risk that could be tolerated, to the organisation in the pursuit of its objectives. This concept is discussed in chapter 10. The risk appetite level should be determined for each of the four categories of security capabilities in Step 1. A key factor in understanding the level of risk is in understanding the Annualised Loss Expectancy (ALE); the estimate of the cost of any loss arising from a failure in cyber security measures within the organisation.

Money Talks

Calculating the Annualised Loss Expectancy

The Annualised Loss Expectancy is derived from the multiplication of the Single Loss Expectancy (SLE) with the Annual Rate of Occurrence (ARO), and is expressed as a formula:

$$\{SLE * ARO = ALE\}$$

Single Loss Expectancy: This is the calculation of the monetary cost of a potential loss if it occurs once. It is calculated using the value of the Assets (AV) at risk, multiplied by the Factor of Exposure (EF) referring to the percentage value of the assets which could be lost. The EF is expressed as a percentage, with 100% representing a loss equalling the total value of the assets at risk. SLE is expressed as a formula:

$$\{AV * EF = SLE\}$$

Annualised Rate of Occurrence: This is the estimate of the likelihood of a specific threat happening within a particular year. The data for the ARO can be taken from historic and industry sources and may be derived as the number of likely occurrences per week or month aggregated into an annual figure representing the number of incidents per year.

Example:

Bob and Alice Ltd. have estimated an ARO for web-based attacks on their infrastructure as being once every week. The ARO is then set as 52; 1 attack per week times 52 weeks per year. The AV is £1000, and the EF is 5%. To calculate the SLE we follow the formula:

$$AV\,(1000) * EF\,(.05) = SLE\,(50)$$

Now that Bob and Alice Ltd. have figures for SLE and ARO, the Annualised Loss Expectancy can be calculated.

$$SLE\,(50) * ARO\,(52) = ALE\,(2600)$$

So, in one year the estimated loss (ALE) to Bob and Alice Ltd. from web-based attacks is £2600.

Now that the potential cost of losses to the organisation is established, the cost and benefits of mitigating the risk must be established. This is often defined as the return on investment where the cost (of countermeasures) is offset against the benefits (reduction in ALE) to derive the countermeasure value. The cost of countermeasures is the sum the of annualised cost of; purchase, implementation, and maintenance.

Calculating the Return on Investment

The Return on Investment (ROI) is a measure used to determine the efficiency of specific expenditure. It is calculated by dividing the ALE with the cost of applying specific countermeasures to mitigate the risk. The benefit or value of the countermeasures is derived from the ALE value, previous to application of any countermeasures, minus the ALE value now that the countermeasures have been applied. The formula used to express this is:

$$\{ALE_{t-1} - ALE_t - CM(c) = CM(v)\}$$

Where ALE_{t-1} expresses ALE previous to the application of any countermeasures, ALE_t is the value now after countermeasures are applied, CM(c) is countermeasure cost, and finally CM(v) is the countermeasures value.

ALE_t is calculated by repeating the ALE formula with updated inputs, that would be the case if the countermeasures were already in place; we should see a reduction in the EF value and the ARO value.

Revisiting Bob and Alice Ltd; let EF equal 2.5% and the ARO is reduced to once every two weeks, giving a value of 26 (52/2). The cost of the proposed countermeasures, including maintenance for a year is £360. Countermeasure Value would be calculated as:

$$AV\,(1000) * EF\,(.025) = SLE\,(25);$$

$$SLE\,(25) * ARO\,(26) = ALE_t\,(1300);$$

$$ALE_{t-1}\ (2600) - ALE_t\ (1300) - CM(c)\ (360) = CM(v)\ (940)$$

Here we can see that the benefit to Bob and Alice Ltd. of employing the countermeasure is £940 reduction in potential losses per year.

Now the return on investment (ROI) can be calculated, given the above values, to show the efficient use of investment funds. This is achieved by dividing the ALE_t and then obtaining a percentage value. It is expressed as:

$$\{(ALE_t / CM(c)) * 100\% = ROI\}$$

Let's ask Bob and Alice Ltd. what the ROI on their investment is.

$$(ALE_t\ [1300] / CM(c)\ [360]) * 100\% = ROI\ (3.61\%)$$

Although calculating the ROI for a loss reduction program may seem redundant, it is a widely used financial tool with which to communicate and justify budgeting requests.

Calculating the value of a network: Metcalfe's Law

Robert Metcalfe is most well known as the developer of the Ethernet protocol, developed the 22nd May 1973, however he is also credited with discovering that the value of a communication network grows exponentially. The law, whilst not a physical law, has been used to estimate the monetary worth

Figure 10.3: Metcalfe's law graph

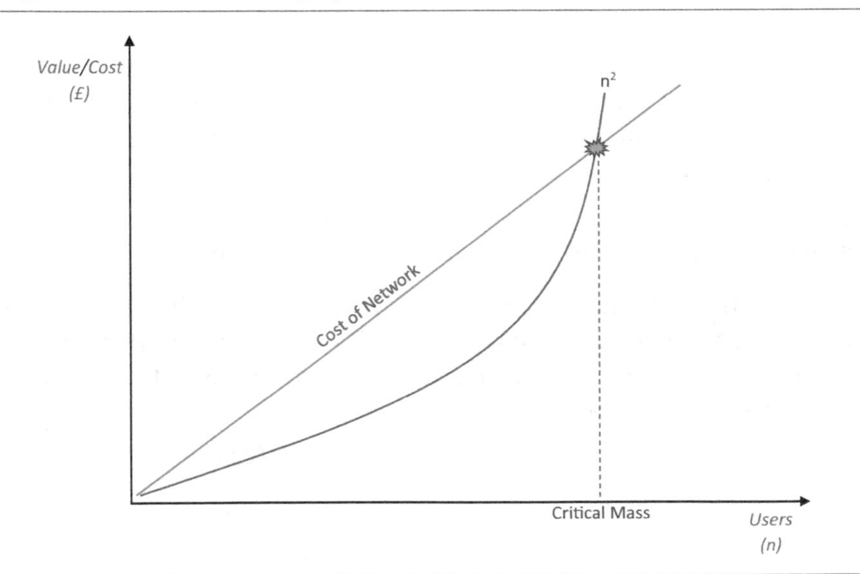

of networked systems, based on the simple premise that where a network has *n* users or nodes, then each can make ($n - 1$) connections within that network. Assuming that each of those connections are of equal value, then the whole network has a value proportional to:

$$\{n\ (n - 1) \text{ or approximately, } n^2\}$$

The application of Metcalfe's Law gives a quadratic function showing that the *cost* of a network is linear, but the *value* of the network is based on the number of possible connections made in that network, given by n^2. The curve shows the point at which the "critical mass" is reached; that is the point in which the network value surpasses the cost of implementing the network (Shown in Figure 10.3).

Effective Communication

Communicating effectively with people outside your field of expertise requires an understanding of the audience; who they are and what do they care about? Often the first point of contact is via email, and there are techniques for enhancing the positive impact of your written communication. Presentations are an even more effective tool with which to communicate information and the need for action, and ultimately achieve the desired outcome. Before beginning work on a presentation, especially the presentation of a budget there are two well-proven theories; on attitudes to risk, and decision making in buying, which should be used to structure the information.

Market for Lemons

The Market for Lemons paper was published in 1970 by George Akerlof, and he was awarded a Nobel Memorial Prize for the insights contained within it. Simply put, in a market where buyers and sellers hold *asymmetric* information; that is the seller knows more about the product than the buyer, then the buyer will choose the lowest cost product. This choice is based on a situation where the buyer perceives the products to be similar, or the same, due to a lack of information about their relative differences.

The example given in Akerlof's paper was used cars; in IT a similar example would be when a buyer with no technical knowledge is comparing two routers: the cheapest one would be chosen. This inability to judge between a "lemon", the poor product, and the better performing product creates a “lemons” market; one where lower quality products (or services) become the market leaders.

In IT and cyber security this is especially the case due to the non-functional nature of many of the requirements: security, reliability, confidentiality, and availability, are outcomes of a well-performing system; not easily measurable until *after* implementation. The important point to take note of, is that it is the buyer’s *lack* of knowledge which drives poor decision making; the antidote to which, is education. The use of visuals (graphs, tables, videos, etcetera), on the relative benefits and costs of each product will help non-technical managers to understand why the proposed product is a better choice for the organisation. When presenting a budget to management for approval, it is imperative that the information on the relative benefits of the preferred option, is explained clearly and simply, with more detailed information available if requested.

Prospect Theory

Forming part of Behavioural Economics, this theory was first developed in 1979, by Daniel Kahneman and Amos Tversky, in an attempt to accurately describe decision making in the face of risk. The fundamental basis of the theory is that people value losses and gains differently: the pain of losing far outweighs the pleasure of gaining.

For example; the pleasure of finding a £10 note in the street is significantly outweighed by the pain of finding two £10 notes, and then losing one of them. Ultimately, the finder still gains £10, however it is the experience of loss which carries the extra weight (described in Figure 10.4): the desire to avoid this feeling is called loss-aversion. In practice this means that people are most likely to make decisions based on potential gains, rather than gains which could expose them to losses.

Understanding Prospect Theory, in the organisational context, will help to structure the narrative focus of presentations into one where the gains are emphasised. Cyber security tends to focus on what could be lost; the “pay now or suffer later” method, but Prospect Theory proves that this message is interpreted as “pay now *and* potentially suffer a loss later”. The very nature of security is loss prevention and mitigation, rather than a guarantee of loss avoidance, which makes the framing of budget requests difficult.

Put simply, managers want to avoid the pain of spending money on security now if they will potentially feel the pain of a security incident later (much like the case of buying a burglar alarm, it tends to only happen when the person has already been robbed). In this case, the gains may be couched in terms of risk mitigation; insurance against the full pain of the potential loss. Below is a graphical representation of prospect theory:

Figure 10.4: Prospect Theory

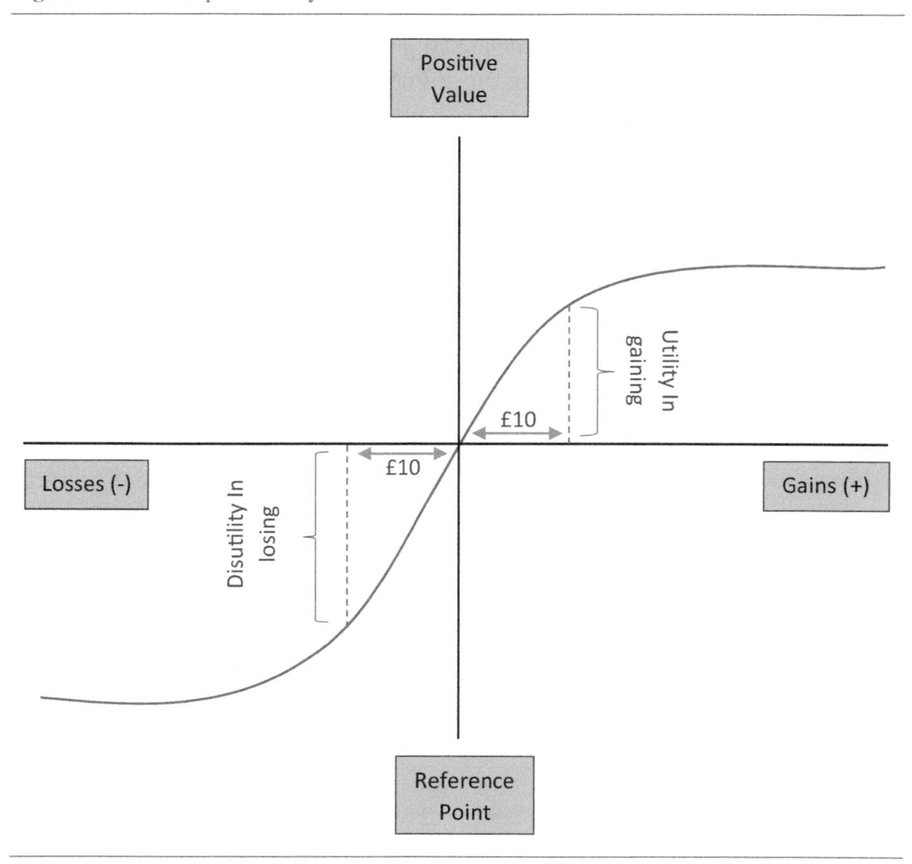

Email

Effective communication requires some planning for the writer and reader to get the most out of the exchange. A well written email will help the writer to achieve the desired outcome, and when that outcome is an increase in the budget allocation, a little time spent in outlining the letter is time well spent. Another important consideration is the role of emails as supporting documentation in compliance audits and investigations. An email is a letter, and letters are written communications between people. Due to the volume of emails exchanged every day, it can be tempting to write a response immediately on receipt, and this often leads to casual and unprofessional communication.

Before you begin:

Think about the outcome; what do you hope to achieve from sending the email? In this chapter we are discussing the business case for cyber security, so the outcome might be along these lines:

- To arrange a meeting with the CFO/CEO to discuss the budget.
- To report on an issue and seek guidance.
- To provide feedback or to follow up on a previous meeting.
- To provide documentation for your actions or of your concerns.

Once the outcome is identified, note down the key points of information you wish to convey. What should the reader clearly understand after reading your email? If there are many points, more than three or four, then it may be better to bullet point them and request a meeting for a detailed discussion.

Be professional:

When drafting an email, especially one to senior management, and most especially one where you are going to ask for money, keep it professional. Think about the layout and readability of your letter; yes, it is an electronic letter. Address it to the relevant person, using their full title, unless otherwise directed remembering that emails are also auditable documents. Keep in mind the fact that others may read the email without your knowledge; be courteous and discreet.

Get to the point:

Business emails are best when they are short and sweet. After the greeting, state the reason for the communication in 150 words, or less, using short, succinct sentences. If more detail is needed, then it can be added in the body of the email. Ideally the reader should know within the first paragraph, why you are contacting them, and what is required from them in return.

Avoid jargon and pomposity:

Be aware of your audience! Is the reader likely to understand acronyms such as SIEM? If they are outside of IT or the C-suite, it is better to err on the side of simplicity and use general terms that your bank manager would understand.

Similarly, overly florid writing full of verbiage is an impediment to the lucid and perspicuous sharing of information, so keep it clear concise and to the point.

We've updated our privacy policy:

The subject header is your email's headline, make it stand out with a clear and positive message; one which will arouse the reader's interest and curiosity, whilst referring to the subject matter of the email below. This is easier to achieve when the header relates directly to the reader's role or department as it highlights what's in it for them. Short titles, which get to the point are more likely to be opened. Unless there is an emergency do not use all capital letters: URGENT should mean exactly that.

Sending the message:

When the email is ready, stop; think about the intended reader. When are they likely to receive and read the email, and is that an opportune time? For example, first thing on Monday morning you send an email to the Chief Financial Officer (CFO) asking for an increase in your budget. In most organisations, Mondays are used for the review and analysis of the previous week's performance and are therefore full of meetings and presentations; Tuesday is usually the same. The correct timing of the delivery of your email is a simple way to improve the chances of getting the outcome you want; between 8pm and midnight has the highest open rates.

Preparing a Presentation

Presentations are a necessary and highly effective tool for the communication of information in person. The function of a presentation is to communicate information and ideas through spoken and visual means, to one or more people, in a clear, logical, and a well-structured way, similar to an essay. Presenting is not teaching, it is sharing, therefore focus on functional rather than technical information when speaking to management. In-person communication is viewed more favourably than its written counterpart when decision making is required, so it is an important part of the budgeting process.

Presentation structure: a beginning, a middle, and an end

- *Introduction:* this section is a brief and informative explanation of who and why. It should introduce the speaker and the topic to be presented and briefly outline the format of the presentation.

- *Body:* here the main points are shared with clear examples that help the audience to understand each point as it relates to them or their department.
- *Conclusion:* end the presentation with a summary of the points covered followed by a concluding remark such as; "In conclusion…", which signal to the audience that the presentation is about to end. An invitation to ask questions can then be offered to the audience.

Step 1: Why?

As explained in the section on email communication, the desired outcome or objective is the basis of why you are creating a presentation; there is a measurable or tangible output that will show the success of the presentation. Ultimately the point is to raise issues and present solutions to those issues by showing how it benefits the organisation through helping it achieve its goals. To do this there must be clear and convincing answers to the following questions:

- *Why* does the organisation or department need this?
- *Why* should the audience care? Whose problems will be addressed?
- *Why* now? What are the current blocks/problems from not doing this?
- *Why* spend? What costs to the organisation in time, money, outputs, etcetera will be reduced?

The point here is to develop a presentation that management will understand and that clearly explains what the proposed outcome will be and why it is necessary for the organisation. The benefits for the organisation are the most important part of answering *why* and help to form the body of the presentation.

Step 2: Bottomline Impact

Once the answers to the question of *why* have been answered it is then necessary to assess the impact on the organisation's bottom line; the profit or loss. Frame the points in the presentation by first stating the issue, in terms of missed opportunities for the business, and illustrate its current impact on the organisation's income or expenditure. Then go on to explain the various solutions available; how the implementation of the proposed solution would reduce the costs. These costs may be in the form of money, time, reputation, regulatory, efficiency, etcetera. The management should have a clear picture of the positive impact this measure would make in *their* part of the organisation as well as the organisation as a whole.

For example; an issue has been identified which could be mitigated through an upgraded firewall. This could be presented in this way:

"Lost productivity due to compromised computers costs approximately 20-man hours per week at an average annual cost of over £10,000. This figure does not include the loss of revenue arising from staff being unable to fulfil their duties. The computers are being compromised, primarily, due to the increased sophistication and frequency of attacks specifically designed to by-pass our current security measures and these attacks are expected to increase. The key security measure identified for the reduction of this cost is the Next-Generation Firewall (NGFW), as it enhances the current protection by detecting and blocking sophisticated attacks, in other words it prevents malicious intrusion.

The following three NGFW's were considered: Brand X, Brand Y, Brand Z. After analysis of the costs and benefits of each, Brand Z has been identified as the best fit for the organisation, based on the higher level of protection offered and the lower ongoing maintenance costs. This will reduce downtime costs to the organisation in total, and especially in the Sales and Marketing departments which have been hardest hit.

Implementation of Brand Z will allow the organisation to improve customer email response time through decreased downtime, and help meet regulatory and contractual obligations more securely."

Keep it broad, to the point, and brief. If management need further details they will ask at the end of the presentation, and many of the questions will have been prepared in Step 1: *Why?* It is also helpful to keep on hand some notes on the more technical and financial details as a backup to memory. Any questions that cannot be answered immediately should be noted down and the answer provided in a follow-up meeting or email.

Step 3: Let the numbers talk

Security is about protection, and protection doesn't generate revenue; it generates *potential, the potential to fulfil objectives and goals without impediment.* It is this potential, for the organisation to run efficiently, to grow more profitably, to transform effortlessly, which must be proved to management.

This is achieved through good quality data and finance metrics which compare the organisation's security posture and spending with those organisations who are similar in size, or market, or both. Historical data can be used to show the impact previous initiatives had on the organisation, using the cost of remediation before the initiative and comparing it to today, as an indicator of the value to the organisation.

Step 4: Be the persuasive expert

Being the expert has its advantages, and as the expert in Information Systems in an organisation, this can be used to enhance the positive reception of the presentation. People are more readily persuaded by the arguments of a trustworthy presenter; one who is honest and open, and experts are considered to be higher in trustworthiness than non-experts. When making a presentation this expertise should be obvious in the confidence with which the points are made, without being overly detailed or technical.

Presentation Visuals

Whether you prefer to use PowerPoint or Prezi, or something else, there a few key considerations that will be the same.

- Do NOT write the script of the presentation on the slides and then read it out loud, verbatim. This is the most often quoted mistake people make and yet it persists. Presenters lose credibility and engagement with the audience and may be perceived as lacking knowledge in their field.
- DO write a headline, similar to the email style, for each slide. This helps orient the audience to the subject at hand, and acts as a visual aid for the presenter. What's written on the slide should trigger the speaker to elaborate on that point.
- DO use images, graphs, and colours to entertain the eyes of the audience and to keep the presentation fresh and engaging. Managers attend many meetings and presentations every day, so an engaging visual display can help them to remember important points.
- DO keep the slides clean; avoid visual clutter which will distract the audience and even confuse them. Slides full of text, crowded graphs, or irrelevant images are counterproductive. Use a font type which is clean and professional in 22 point or larger, and use the same font throughout the presentation.

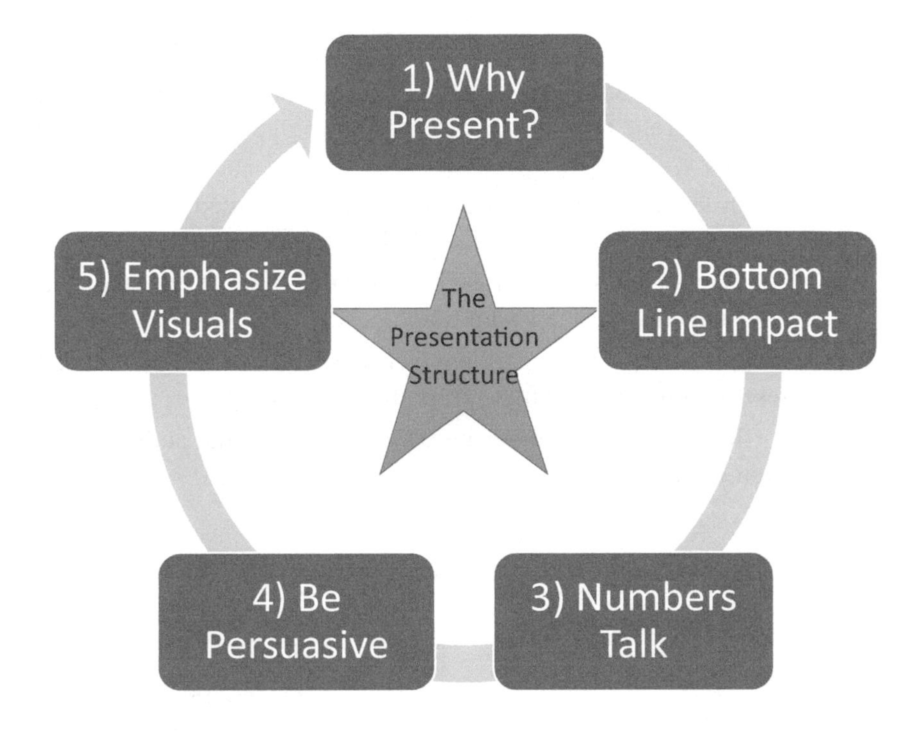

The 10/20/30 rule of presentation slides

A simple framework for the timing and structure of presentations is the 10/20/30 Rule, attributed to Guy Kawasaki, one of Apple Corporation's earliest employees. The rule is an effective starting point because it is easy to remember and simple to execute. There are three key parts:

- There should be no more than ten slides.
- The presentation should last 20 minutes; plus, additional time for questions.
- The font size should be 30 point and larger.

Where longer presentations are required, the rule of one slide for every two minutes is a reasonable extrapolation of the 10/20/30 rule. Each slide has a specific role within the presentation story arc; the scene setting, the main body of the story, and the ending.

A simple ten slide presentation would follow these lines:

Slide 1: Orients the audience to the subject of the presentation. It should include the title of the presentation, the presenter's name, and role or subject expertise.

Slide 2: High-level overview of the issues being presented. This should provide sub-headings which the presenter will elaborate.

Slides 3–8: Supporting slides for each of the key points. Again, these should provide a visual cue to the relevant point.

Slide 9: Supplementary information which was not covered in the main body of the presentation but is useful or important to share.

Slide 10: Closing image or message which signals to the audience that the presentation is over.

Chapter Review

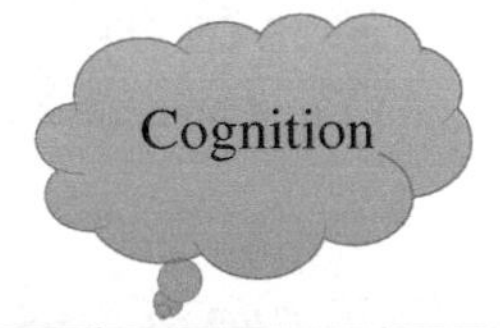

Summary of Key Points

1) Budgeting Process for cyber security.
2) Mapping out a budget and what to discuss with senior management.
3) Money is the language of finance and so too is a well-developed budget for IT.
4) Calculating the ALE and ROI of implementing cyber security protocols and tools.
5) Discussion on Metcalfe's law and the value of a network.
6) A look at prospect theory and the market for lemons.
7) The use case for effective communication within the IT department.

- What would you add to an IT budget and how would you communicate that to management.
- How could you best frame the network value of the organisation?
- Think about other creative ways you could present the business case for cyber security to senior management.

References

Akerlof, G. (1970). The Market for Lemons: Quality Uncertainty and the Market Mechanism. *The Quarterly Journal of Economics, 84*(3), 488–500.

Certified Institute of Management Accountants. (2005). *Budgeting*. Retrieved from CIMA: Topic Gateway Series No. 27: https://web.archive.org/web/20130810055251/http://www.cimaglobal.com/Documents/ImportedDocuments/cig_tg_budgeting_mar08.pdf

DEBORAH SMITH. (2002, 12). Psychologist wins Nobel Prize. *Monitor on Psychology*, p. 22.

Experian Marketing Services. (2012). *Quarterly Email Benchmark Study.* Experian Information Solutions Inc, Schaumburg.

Gallaher, Michael P; Rowe, Brent R; Rogozhin, Alex V; Link, A. (2006). *ECONOMIC ANALYSIS OF CYBER SECURITY.* USAF Research Triangle Institute , Rome, New York.

Gordon, l; Loeb, M; Lucyshyn, W; Zhou, L. (2015). Externalities and the Magnitude of Cyber Security Underinvestment by Private Sector Firms: A Modification of the Gordon-Loeb Model. *Journal of Information Security, 6*, 24–30.

Hendler, J., & Golbeck, J. (2008). Metcalfe's law, Web 2.0, and the Semantic Web. *Web Semantics, 6*(1), 14–20.

Hershberger, P. (2016). *Planning, Budgeting and Communicating the Critical Security Controls Implementation.* SANS Institute.

IEEE Spectrum Magazine. (2006, 6). *Metcalfe's Law Is Wrong*. Retrieved from Physics.org: https://phys.org/news/2006-06-metcalfe-law-wrong.html

Inland Revenue Service (IRS). (2017). Cybersecurity Dashboard on a Shoestring Budget. 1–22. Computer Security Resource Centre.

Kahneman, Daniel and Tversky, A. (1979). Prospect Theory: An Analysis of Decision under Risk. *Econometrica, 47*(2), 263–91.

Kawasaki, G. (2018). *The 10/20/30 Rule of PowerPoint*. Retrieved from guykawasaki.com: https://guykawasaki.com/the_102030_rule/

Kemp, Nenagh; Grieve, R. (2014). Face-to-face or face-to-screen? Undergraduates' opinions and test performance in classroom vs. online learning. *Frontiers in Psychology, 5*, 1278.

Klucharev V, Smidts A, F. (2008). Brain mechanisms of persuasion: how "expert power" modulates memory and attitudes. *Social Cognitive and Affective Neuroscience, 3*(4), 353–366.

Lindsey, M. (2010). *A Method for Estimating the Financial Impact of Cyber Information Security Breaches Utilizing the Common Vulnerability Scoring System and Annual Loss Expectancy.* Faculty of the Graduate School of The University of Kansas .

Mersinas, Konstantinos; Hartig, Bjoern; Martin, Keith M.; Seltzer, A. (2016). Are information security professionals expected value maximizers?: An experiment and survey based test. *Journal of Cybersecurity, 2*(1), 57–70.

Metcalfe, B. (2013). Metcalfe's Law after 40 Years of Ethernet. *Computer, 46*(12), 26–31.

Odlyzko, A., & Tilly, B. (2005). A refutation of Metcalfe's Law and a better estimate for the value of networks and network interconnections. *Technology* (December), 1–11.

O'Hara, C. (2014, 11). How to Improve Your Business Writing. *Harvard Business Review.*

Priester, Joseph R; Petty, R. (2003). The Influence of Spokesperson Trustworthiness on Message Elaboration, Attitude Strength, PRIESTER AND PETTY ENDORSER TRUSTWORTHINESS and Advertising Effectiveness. *JOURNAL OF CONSUMER PSYCHOLOGY, 13*(4), 408–421.

Radziwill, N. M. & Benton, M. (2017). Cybersecurity Cost of Quality: Managing the Costs of Cybersecurity Risk Management. *Software Quality Professional, 19*(3).

Rimmer, D. (2009). *NAVIGATING DIFFICULT TIMES.* London School of Economics. London: Hewlett Packard.

Roghanizad, M. Mahdi; Bohns, V. (2017). Ask in person: You're less persuasive than you think over email. *Journal of Experimental Social Psychology, 69*(3), 223–226.

SHIMAMOTO, D. (2012). A strategic approach to IT budgeting. *Journal of Accountancy* (3).

Tarala, J. (2015). Pragmatic Metrics for Building Security Dashboards. *RSA Conference 2015*, (pp. 1–30). San Francisco.

Tomlinson, P. (2016, 11). Don't open the door to cybercrime. *Accounting Business.*

Index

About the Authors

Connor Fowler

Connor Fowler an associate of the Cyber Academy (Edinburgh Napier University) is a graduate of Economics from Heriot-Watt University. He has international experience in risk management, working on alternative energy projects in Mumbai and gas and oil projects in Dubai. His experience in risk management coupled with specialised knowledge on the GDPR allows him to bring business reality and academic theory together.

He is currently working on a project researching the impact of distributed ledger technologies on economic principles, the project has the working title; *Encryptonomics: Living in a Decentralised Economy.* This hypothesis investigates existing economic theories of money and information relevant to the disruptive capacity of distributed ledger technologies, such as cryptocurrencies.

Antoni Gobeo

An associate of The Cyber Academy, Antoni first developed an interest in cryptocurrencies and distributed ledger technology whilst on the LLB course at Edinburgh Napier University. A natural consequence of which was a growing understanding of the importance of data protection law in supporting the human right to privacy. Extensive business experience in design and supply chain management enhances her ability to apply academic knowledge to achieve practical real-world solutions; for business and research.

Currently working on several projects, Antoni's main focus is on developing educational tools for all age groups, on the topic of data protection, and researching the commercial applications of blockchain technology for enhanced food security. This latter focus has multiple areas for research including; sustainable urban agriculture (including vertical farming), supply chain management, full-cycle food provenance, and the real-time measurement of economic impacts arising from urban Agritech.

William Buchanan

William (Bill) Buchanan is a Professor in the School of Computing at Edinburgh Napier University, and a Fellow of the BCS, and the IET, and a Principle Fellow of the HEA. He was appointed an Officer of the Order of the British Empire (OBE) in the 2017 Birthday Honours for services to cyber security. Bill recently received an "Outstanding Contribution to Knowledge Exchange" award at the Scottish Knowledge Exchange awards, and was the Cyber Evangelist of the Year in 2016. Currently he leads the Centre for Distributed Computing, Networks, and Security and The Cyber Academy (thecyberacademy.org). His research has led to three successful spin-out companies, and has developed a number of patents. He has been named as one of the Top 100 people for Technology in Scotland for every year since 2012, and was also included in the FutureScot "Top 50 Scottish Tech People Who Are Changing The World".